I0160570

NATIVE AMERICAN
PLACENAMES
OF THE SOUTHWEST

NATIVE AMERICAN

PLACENAMES

OF THE SOUTHWEST

A HANDBOOK FOR TRAVELERS

William Bright

Edited and with an
Introduction by
Alice Anderton and Sean O'Neill

UNIVERSITY OF OKLAHOMA PRESS : NORMAN

Library of Congress Cataloging-in-Publication Data

Bright, William.
Native American placenames of the Southwest : a handbook for travelers / William
Bright ; edited and with an introduction by Alice Anderton and Sean O'Neill.
   p.   cm.
Includes bibliographical references.
ISBN 978-0-8061-2444-5 (pbk. : alk. paper)
1. Names, Geographical—Southwest, Old. 2. Names, Geographical—Southwest,
New. 3. Names, Indian—Southwest, Old. 4. Names, Indian—Southwest, New.
5. Southwest, Old—History, Local. 6. Southwest, New—History, Local.
I. Anderton, Alice, 1949–
II. O'Neill, Sean, 1969–
III. Title.
F396.B85 2013
979—dc23   2012020080

The paper in this book meets the guidelines for permanence and durability of the
Committee on Production Guidelines for Book Longevity of the Council on Library
Resources, Inc. ∞

Copyright © 2013 by the University of Oklahoma Press, Norman, Publishing
Division of the University. Manufactured in the U.S.A.

All rights reserved. No part of this publication may be reproduced, stored in
a retrieval system, or transmitted, in any form or by any means, electronic,
mechanical, photocopying, recording, or otherwise—except as permitted
under Section 107 or 108 of the United States Copyright Act—without the
prior permission of the University of Oklahoma Press. To request permission to
reproduce selections from this book, write to Permissions, University of Oklahoma
Press, 2800 Venture Drive, Norman OK 73069, or email rights.oupress@ou.edu.

CONTENTS

MAPS

AUTHOR'S INTRODUCTION

This book is aimed at the general reader who may be curious about the origins of placenames in the American Southwest—Oklahoma, Texas, New Mexico, and Arizona—an area that uniquely brings together American Indian, Mexican Spanish, and Anglo-American history. In particular, I hope that visitors driving through the area will tuck the book into the glove compartments of their cars for ready reference as they pass through what was, and remains, Indian country.

This book represents a shortening and adaptation, for a general readership, of information previously published in my larger work *Native American Placenames of the United States* (University of Oklahoma Press, 2004). More detailed linguistic analyses are given there, as well as full information about my published and unpublished sources for that volume. The most important published sources are listed at the end of this volume.

The origins of many Indian placenames are still unclear or controversial. Please send clarifications or corrections to the editors at the University of Oklahoma Press.

EDITORS' INTRODUCTION

William Bright, the author of this book, was an eminent linguist and anthropologist who earlier published several other books about American placenames—most famously his 2004 *Native American Placenames of the United States*—as well as scores of highly regarded works on and bibliographies of Native American and East Asian languages and other scholarly topics. Professor Bright passed away in 2006, just before completing his work on this project. In 2008, we began working with the draft he left behind, filling in gaps in the research and revising the entries to make them more consistent. To honor our colleague and mentor, we agreed to prepare the manuscript for publication, as we think Bill would have wanted, so that the public might benefit from the enormous amount of useful research it represents.

This guide is derived partially from Bright's 2004 book, which itself was based on a wide range of sources, including the Geographical Names Information System (GNIS), the digital database of placenames available from the U.S. Board on Geographic Names; existing placename dictionaries of the United States as a whole; regional placenames dictionaries such as Byrd Howell Granger's *Arizona's Names* ; and regional Native American placename dictionaries such as Laurance Linford's *Navajo Places: History, Legend, Landscape.*

What Bright intended to include were names that people are likely to encounter in books, on maps, or on the land and about which they feel curious.

In compiling the present volume, Bright added some new entries from the four southwestern states, and we added a few more, drawing from state maps, personal knowledge, and our own linguistic consultants. As editors we took on the task of filling in gaps in information and rewording and reformatting the manuscript in a few ways to shorten it and make it more user-friendly for nonspecialists. Our primary purpose in the way we present the entries is to provide travelers with a handy resource in which to look up the meanings and histories of placenames they come across on the road— making their travel, we hope, more interesting. Where possible, we have included pronunciations, although some of the placenames are not widely circulated today and so have no standard English pronunciation. We have minimized academic terminology and abbreviations and kept the volume small, so that it will fit into a glove compartment, purse, or travel bag. We have also added maps, a list of Native languages of the area and the linguistic families to which they belong, and contact information for American Indian tribal groups in the Southwest.

The scope of this book is the American Southwest, defined as the present states of Arizona, New Mexico, Oklahoma, and Texas. Before Oklahoma was a state, tribes in what was then called Indian Territory each had their own territorial boundaries, which are sometimes approximated today by tribal service boundaries within the state. Phrases in the placename listings such as "in the historic Creek Nation" and "in the traditional Cherokee Nation" refer to these old boundaries. We include cities, towns, counties, and states, as well as a few former townsites, parks, and geographical features such as lakes and creeks.

The places listed in the book have names of three main kinds:

1. Names drawn directly from Indian languages, such as **Oklahoma**, which means 'Indians' in Choctaw, and **Tucson,** a name in the O'odham language of southern Arizona.

2. Names derived from Indian names by translation or by adaptation to English, such as **Gossip Hills**, New Mexico, which is translated from the Navajo word for 'gossip', and **Texas**, from the Caddo word *tayshah,* 'friend', by way of Spanish, with the pronunciation adapted to Spanish and then to English.

3. Names that come from Indian personal names, such as **Tecumseh**, Oklahoma. Sometimes the source is not the first one that comes to mind. **Roman Nose**, a state park in Oklahoma, is named not after Romans but after Henry C. Roman Nose, a Cheyenne man. **Washington**, Oklahoma, is named after a Caddo man, not the U.S. president.

Of course many Native American placenames in the Southwest are used only among speakers of Native languages, not by the general public, and do not appear on maps and road signs. An example is Comanche **Piaroya** (literally 'big mountain'), which is called Mount Scott in English. No attempt has been made to list these, because travelers are unlikely to encounter them.

Where Placenames Come From

As far as we know, people have always given names to places. One common way of naming places is to refer to the look of the place or to its most salient geographical feature. Names that refer to geographical features are called *toponyms.* In English, this leads to names such as Smoky Mountains and Crater Lake. An example in a Native language is **Ali Wua Pass**, Arizona, from an O'odham name meaning 'little pond'.

Another way of naming refers to the people who live or formerly lived at a place or to famous or revered persons from the area. This mode results in names such as **Comanche** and **Geronimo,** both in Oklahoma, and **Navajo County**, Arizona. The name is often combined with a toponym or with a word or word part indicating a settlement, forming names such as **Webber's Falls** and Plunkettville, both in Oklahoma. Names of local plants and animals are sometimes used in placenames, too, as in **Pecan Creek**, Oklahoma, and Skunk Creek, Texas.

Sometimes the name of a place commemorates an event, a myth, or a story associated with the place, as in the cases of Massacre, New Mexico, and **Tat Momoli**, Arizona, the latter named for a footrace. Counties are often named after the county seat, as in **Atoka** and **Atoka County**, Oklahoma, or vice versa. Sometimes places are named after nearby states, such as **Texas County**, Oklahoma. Existing names can be blended to produce new ones; **Lake Texhoma**, for instance, combines *Texas* and *Oklahoma.*

Less common ways of naming places include using the name of a famous or revered person who actually had little or nothing to do with the place, as in the example of Jefferson, Texas, which has no direct connection to Thomas Jefferson. In other cases, a now obscure person was connected with a settlement, an example being Rotan, Texas, which managers of the railroad that passed through the town named after an investor. Similarly, places are sometimes named for someone who had a personal connection to the person who did the naming. In Oklahoma, where towns sprang up almost too fast to name, many towns, such as Ada, were named for the postmaster, his wife, or his daughter. Other names come from popular literature or cultural icons, such as **Hiawatha**, Oklahoma. An unusual means of naming, which we find amusing, is to reverse some of the sounds, as in **Amabala**, Oklahoma, based on *Alabama.*

In areas with a history of invasion or colonialism, one often finds that old names given by the original inhabitants survive in the usage of the newcomers, who simply adopted the existing name rather than make up a new one. Linguists usually call this practice "borrowing," although "adoption" or "requisition" might be more accurate, because the term cannot, of course, actually be returned. Thus we get Spanish speakers using the name *Mexico,* which was originally an Aztec name referring to the inhabitants of the area around present-day Mexico City, and French speakers in Louisiana using *Natchitoches,* which was originally the name of a Caddo Indian group.

The pronunciation of Native names is often adapted to the language of those who adopt it, who usually only approximate the original. The pronunciation of *Natchitoches,* Louisiana, for example, is not the same as the pronunciation of the original Caddo placename. Sometimes existing placenames are translated into the new language, as in the case of **Echo Cave,** Arizona, whose name was translated from a Navajo expression meaning 'rock under which there is an echo'.

In another kind of borrowing, a "transfer name"—a placename from another area—may be used in a new location, often by immigrants who want to maintain a connection with their homeland. Examples are New York, after York, England; Boise City, Oklahoma, after Boise, Idaho; and **Arbeca Creek,** Oklahoma, after *apehkv,* a Muskogee tribal town in Georgia.

Representing all these ways of naming and more, the placenames one sees on road signs and maps in the southwestern United States reveal aspects, sometimes no longer obvious, of local history, language contacts, attitudes, and geographical features. Placenames often reflect the history of an area—for instance, the important role played by water sources, other resources, and aspects of the land, or the occurrence of battles and other events. Contacts between cultures are made evident in, for example, the prevalence of Spanish placenames in Arizona, New Mexico, and Texas and the appearance of French placenames in eastern Oklahoma. Personal names tell us which noteworthy individuals lived in the area or were held in high regard. Names derived from Native placenames often tell us which tribes live or have lived in the area. In some cases the placename itself is the only hint of the place's history, so understanding the name is important to historians of the area as well as to the interested public.

How to Use This Guide

Each placename entry in this book begins with the name itself, followed in parentheses by the location of the place—that is, the two-letter postal abbreviation for the state and then the name of the county. The pronunciation of the name, where known, follows, between slash marks. The etymology, or historical derivation of the name from the Native expression, follows that. Sometimes there is further discussion of the name's history, especially if it is particularly unusual or interesting, as in the following example:

CHICORICA Creek (NM, Colfax Co.) \chik uh REE kuh\. Perhaps from a Plains Apache term for 'turkey', equivalent to Navajo *tsídii łíkizh* 'spotted bird' (lit. 'bird-spotted'). This may have become Spanish *chiti-riqui,* changed by assimilation to *chiqui-riqui* and then reinterpreted through folk etymology to *chico rico* 'rich child'. . . .

To interpret the pronunciation of a placename, consult the guide to English pronunciation that follows this introduction. To interpret the pronunciation of a Native American expression cited as the source of a placename, consult the section "Guide to the Pronunciation of Native Words," also following the introduction, or relevant works in the references section.

Sometimes, more than one pronunciation of the placename is given, as in the case of **Tecumseh**, Oklahoma, which is locally pronounced /tuh KUM see/ by some people and /tuh KUM suh/ by others. In these cases, we give all the acceptable pronunciations of which we are aware.

For most entries we give the pronunciation or pronunciations of the placename as they are said by English speakers. We have made every effort to give the *local* pronunciation. For example, we tell you that **Miami**, the town in Oklahoma, is pronounced /mie AM uh /, unlike Miami, the city in Florida, which is pronounced /mie AM ee/. If we were not personally familiar with a place, we consulted knowledgeable local people, usually at libraries and other public institutions. In cases where such information was unclear or unavailable to us, we have omitted it.

When we give the placename in the Native language, we use the standard spelling for the language when there is one, and otherwise we follow the phonetic representation used by scholars of the language.

Sometimes a name or variations of it occur in other places as well, and in those cases we include a reference to the related name, such as "A related placename is . . ." or "An alternate spelling is . . ." When a separate entry for the related term exists, it is written in all capital letters, and you may find additional information of interest under the cross-referenced entry.

In looking up a placename, ignore small words such as "El," "Los," and "The," as well as hyphens and apostrophes. The name **El Guique,** for example, is listed as **GUIQUE, EL.** Also ignore generic names such as *lake, arroyo,* and *cañada.* The name **Arroyo Zacatoso,** for instance, is listed as **ZACATOSO, Arroyo.**

To learn more about a language that is mentioned in a placename entry, please consult the language list that makes up Appendix 1. It will give you the language family to which the language belongs. Also check the selected references at the end of the book for further reading.

Special Southwestern Terms

Those who are new to the Southwest may be unfamiliar with a few terms used frequently in its placenames:

arroyo. A seasonal stream or its dry bed.

banco. A levee.

barranca. A narrow, winding river gorge.

bayou. A sluggish stream, typically found in Louisiana but also occurring in Texas and Oklahoma.

cañada. A ravine or small canyon or creek.

canyon. A deep river valley with steep sides of precipitous cliffs.

mesa. A geographical form emblematic of the American Southwest, especially New Mexico. In effect a truncated hill, it is a substantial area of higher elevation, with nearly vertical sides and a flat rather than rounded top. The First, Second, and Third Mesas on the Hopi Reservation are home to the Hopi and Arizona Tewa peoples.

pueblo. The Spanish word for 'village, town', which came to be used to refer to the compact Indian settlements in Arizona and New Mexico and the people who live in them—the Hopi, Zuni, Keres, and Tanoan peoples.

Special Linguistic Terms

Although we have attempted to use in this book everyday language understandable by anyone, we have retained a few useful terms that are common to many placename guides. The following list explains a few of these that may be helpful to readers who are new to placename terminology:

blend. A blend is a combination of two or more words, usually the first part of one word and the last part of another, as in *smog,* containing the first part of *smoke* and the last part of *fog.* A good Southwest placename example is **Texhoma,** which is made up of the first part of *Texas* and the last part of *Oklahoma.*

borrowed. A name or other word is said to have been "borrowed" from another language when speakers of the borrowing language decide to use it, or an approximation of it, for a new thing, concept, or place, rather than inventing a new word of their own. Of course "borrowed" terms, unlike borrowed goods, cannot be given back, but they shed some light on contact between speakers of different languages in an area, including contact between immigrant Americans and Native peoples. Examples are **Tucson**, Arizona, borrowed from O'odham, and **Texas**, borrowed from Caddo by way of Spanish.

diminutive. A diminutive is a form of a word that denotes a smaller version of the thing being named. An example is *isleta*, as in **Isleta,** New Mexico, meaning 'small island' or 'islet', from *isla* 'island'.

etymology. The etymology of a word is its linguistic history, which usually includes the language from which it originally came, its original meaning, perhaps its original spelling or pronunciation, any related story or explanation that might be known for it, and any other known information about the word's path as it found its way into current usage. For example, **Tennessee Colony**, Texas, was named after either the state of Tennessee or the Tennessee River, both of which in turn were named after a Cherokee village. Most often in placenames, the reason the old word or phrase forms the basis of the modern placename is that it describes a geographical feature in the area, an event that took place there, or a person or a people who lived or live there. In a few cases the history is not so logical. Perhaps someone knew the name from another place and chose it because he or she liked the sound or feel of it for some reason (see *transfer name*, below). In some cases placenames are applied because of misunderstandings (see *folk etymology*, below). In this book we give all such information of which we are aware.

folk etymology. This term refers to a popular but inaccurate understanding of a word or placename that causes speakers to distort it. An example is **Salina,** Arizona, a name that comes from Navajo *tséláni* 'many rocks', which non-Navajo speakers have reinterpreted as Spanish *salina* 'salt flat'.

literally. In modern everyday usage, *literally* has come to mean something vaguely like "really," as in the remark "He literally blew my mind." Linguists and translators use it in a more technical sense to distinguish the word-by-word breakdown of an idiomatic phrase or a complex word from its meaning as used. Many examples can be found in everyday English idioms such as "blow one's mind." That phrase is generally understood as and used to mean "astound," although its literal meaning—the meaning one would arrive at by examining the meaning of each of the words and combining them as if they were an ordinary phrase—is something closer to "explode one's brain." Translations can be either literal or nonliteral. For example, *okla homma,* the Choctaw expression from which the state name **Oklahoma** is derived, literally (by translation of each of its parts in order) means 'people red', but in everyday usage the phrase is the Choctaw way of talking about American Indian people.

toponym. A toponym is the name of a geographical feature such as a lake, mountain, waterfall, or canyon, often used as the name of a nearby place. Some writers use *toponym* for any placename, regardless of source.

transfer name. This is a name copied from some other place with the same name. Examples are **Agawam** and **Etowah Church**, both in Oklahoma. Immigrants to an area typically transfer a name because a place reminds them of another place they know or out of nostalgia for their home.

A Note on Translations

Sometimes word order in a Native language expression that is the historical source of a placename differs from the word order the expression would take in English. This kind of difference, which must be accommodated in translation, is common because of the different patterns that are normal in different languages. For instance, adjectives commonly follow the word they modify in Spanish, whereas the opposite order is usual in English. Thus, a good translation of *casa blanca* is 'white house', even though the literal translation (taken in the order of the Spanish) would be 'house white'.

Following the principle that a good translation sounds normal in the language into which one is translating—not broken, stilted, or foreign— we have translated placenames into normal-sounding English equivalents of their meanings. Where appropriate, we note the original word order as part of the name's literal meaning. For example, we translate *okla homma,* the Choctaw source of **Oklahoma**, as 'Indian', not as its literal form 'people red'—let alone "land of the red man" or any of the other popular but fanciful translations so frequently offered. Similarly, **Dibe Chaa Valley,** literally 'sheep-big' valley, is translated as 'big sheep valley'.

Additional Resources Included in This Guide

Maps. Maps of each of the four southwestern states, showing the locations of counties and major cities, can be found on pages xxvii–xxx.

Information about Language Families. Linguists group the languages of the world into "families," just as biologists group the plants and animals of the world into families, on the basis of common descent. Language families are sometimes represented in branching diagrams, or "family trees," to show how they are related to one another. The Romance languages, for example, descended from an ancestral language that was (roughly) Latin, which in turn descended (as did the Germanic languages and many others) from an even older ancestral language called Proto-Indo-European. The Uto-Aztecan languages spoken by Native peoples in the present-day United States descended from an ancestral language called Proto-Uto-Aztecan. These groupings are based on observed similarities of vocabulary and structure. Appendix 1 lists the languages mentioned in this guide and their language families.

Tribal Contact Information. A list of tribal headquarters in each of the four southwestern states, with addresses, telephone numbers, and World Wide Web addresses, is given in Appendix 2.

Selected References. This list includes a few major reference works on Native American languages of the United States, especially those of the Southwest. Readers who are interested in more detailed linguistic analyses and sources may wish to consult William Bright's earlier volume, *Native American Placenames of the United States.*

GUIDE TO ENGLISH PRONUNCIATION

The following is a key to the system William Bright used to represent the English pronunciations of the placenames in this guide, including the corresponding spelling of the sound in the International Phonetic Alphabet (IPA).

Bright's System	IPA	As in (approximate sound in English words)
\a\	[æ]	cap, bad, act
\ah\	[a]	father, spa, far
\ay\	[eɪ]	day, wait, cape
\au\	[aʊ]	house, cow, down, out
\aw\	[ɔ]	caught (for those who pronounce it differently from cot)
\ch\	[tʃ]	child, church, cheese
\e\	[ɛ]	set, red, left
\ee\	[i]	tree, east, eve
\g\	[g]	go, give, gag
\i\	[ɪ]	it, bid, ink
\ie\	[aɪ]	pie, high
\ng\	[ŋ]	sing, long
\r\	[ɹ]	roar, road, heart, dear
\sh\	[ʃ]	shine, hash, dish
\oo\	[u]	boot, soon, coop
\oh\	[oʊ]	hone, bone, phone
\oy\	[oɪ]	boy, joy
\u\	[ʊ]	put, foot, book
\uh\	[ə]	a as in English sofa
\ur\	[ɚ]	fur
\zh\	[ʒ]	azure

The following are pronounced as in ordinary English: b, d, f, j, m, p, s, t, v, y, z. A capitalized syllable is stressed, as in Texas /TEK suhs/.

GUIDE TO THE PRONUNCIATION OF NATIVE WORDS

The following symbols are among those used in the most widely represented languages in this volume.

Symbol	Pronunciation
[a]	As in English *spa*, Spanish *padre*
[æ]	As in English *bad, cat*
[c]	May vary from *ts*, as in English *gutsy*, to *ch*, as in English *church*
[č]	As in English *church*
[ḍ]	A sound like *d*, but with the tip of the tongue slightly curled back
[ð]	Like the *th* in English *the, whether*
[e]	As in English or French *sauté*, Spanish *bebe*
[ə]	Like the *a* in English *sofa*
[ɣ]	A sound like *g*, but made with some turbulence and without closing off the airstream, as in Spanish *lago*
[i]	As in English *machine*, Spanish *sí*
[ɪ]	As in English *it, bid*
[ɨ]	A "mid-central" or "high central" vowel, impressionistically between [i], [ə], [ʊ] (see elsewhere on chart)
[ɬ]	A sound like *l*, but with a whispery, voiceless sound
[ñ]	As in Spanish *señor*
[ŋ]	Like English *ng* in *sing* or *n* in *sinker*
[o]	As in English *solo*
[ɔ]	Like the vowel of English *caught*, when distinct from *cot*
[q]	A sound like *k*, but formed farther back in the mouth
[r]	In many languages indicates a tapped *r*, as in Spanish *toro*, without the darker sound of English *r* as in *roar*
[š]	The *sh* sound in English *ship, shore*
[ṣ]	A sound like *s*, but formed with the tip of the tongue curled slightly back
[ṭ]	A sound like *t*, but formed with the tip of the tongue curled slightly back
[θ]	Like English *th* in *thin, theater*

[u]	The vowel sound in English *dune, boot, soup*
[x]	Like a k, but made without closing off the airstream; like the *j* in Spanish *José* or the *ch* in German *Bach*
[ž]	Like the middle consonant in English *azure, vision, leisure.*
[ʒ]	Like English *dz* in *adze.*

The ['] apostrophe, when written after some consonants, indicates that the sound is glottalized—that is, pronounced with a sharp, popping sound and brief closure of the vocal cords.

Accent marks indicate pitch. Usually, the acute accent [á] marks strong accent or high pitch, the grave accent [à] marks low pitch, and the circumflex accent [â] marks falling pitch.

A hook below a vowel symbol indicates that the vowel is nasalized. Thus [ǫ] is nasalized, as in French *mouton* 'sheep'.

COMMENTS ON SPELLING SYSTEMS

Some Native American languages employ practical spelling systems that use additional symbols or use symbols shown in the preceding chart with other pronunciations. These are given below.

Aztec (Nahuatl)

Vowels and *y* are pronounced as in Spanish.

ll is a long *l* sound, not *y* or *ly* as in Spanish

c is pronounced like *k*

qu is pronounced like *k*

tl is a combination of *t* and *l* ("voiceless lateral affricate"), with the *l* component being breathy rather than resonant as in English *bottle*

tz is pronounced like the *ts* in English *cats*

x is pronounced like English *sh*

z is pronounced like *s*

Cherokee

v represents IPA [ʌ̃], the vowel [ʌ] of English *bud*, with added nasalization

Comanche

b is pronounced like Spanish *b* or *v* between vowels—that is, with the lips not fully closed

r is pronounced as in Spanish, as a tongue tap against the ridge behind the teeth

ʉ, called "barred *u*," is a central vowel, similar to the second vowels of English *sofa* and *mannequin*

Underlined vowels are whispered.

Kiowa

b, d, g, h, l, m, n, s, w, y and *z* are pronounced as in English

p	an aspirated (breathy) *p*, as in English *pie*
t	an aspirated (breathy) *t*, as in English *tie*
k	an aspirated (breathy) *k*, as in English *Kate*
f	an aspirated (less breathy) *p*, as in English *spy*
j	an aspirated (less breathy) *t*, as in English *sty*
c	an aspirated (less breathy) *k*, as in English *skate*
v	an ejective *p* (made like *f* but with a little catch after it)
x	an ejective *t* (made like *j* but with a little catch after it)
q	an ejective *k* (made like *c* but with a little catch after it)
au	equivalent to IPA [ɔ]
´	rising tone
`	falling tone
^	rising then falling tone

Vowels followed by : are held a little longer than other vowels.

Underlined vowels are nasalized.

Muskogee (Creek)

Following Martin and Mauldin's *A Dictionary of Creek/Muskogee*, words are given both in the practical spelling system and in phonemic transcription (for example, yvhv /yʌhá/ 'wolf'). Special usages of the practical system are as follows:

c	IPA /č/ as in English *church*
e	IPA /i/ like the vowel in English *if*
ē	IPA /i:/, like the vowel of English *feet*
ī	formerly pronounced as /ai/, like the vowel of English *pine*; now pronounced as /ei/, as in English *hey*
r	IPA /ɬ/, "a voiceless *l*" (see [ɬ] above)
ue	IPA /oi/, as in English *boy*
v	a short /a/, pronounced [ʌ], as in English *bud*

Navajo

Following Young and Morgan's *The Navajo Language*, the sequence *gh* is used to represent [ɣ].

O'odham (Pima, Papago)

In the official spelling, the symbol ĭ represents voiceless *i* [i̥], and *c* is pronounced as *ch* [č].

Osage

The vowel *ü* is pronounced like the German *ü* in *früh* 'early'.

Shoshoni

The letter *e* is used to represent IPA [ɨ] (see previous chart).

ABBREVIATIONS USED IN THE PLACENAME ENTRIES

AL	Alabama
AZ	Arizona
CA	California
lit.	literally
NM	New Mexico
OK	Oklahoma
pl.	plural
sing.	singular
spp.	species
TX	Texas

MAPS

Arizona and Its Counties

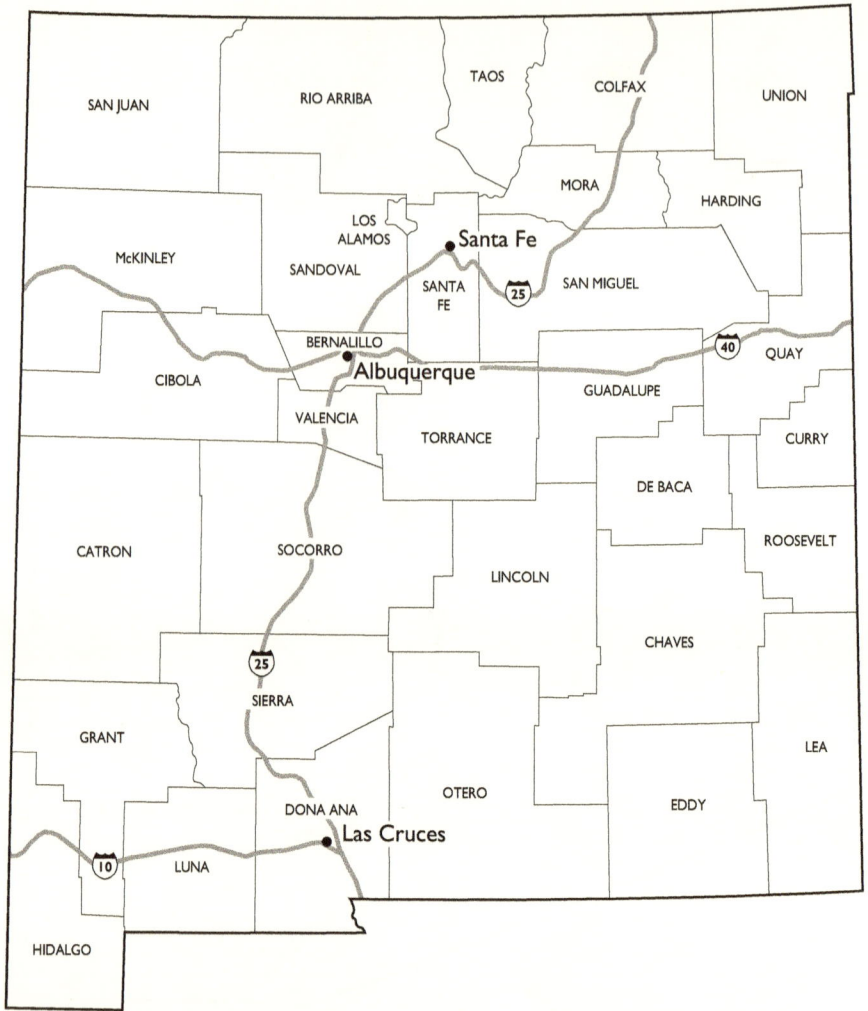

SAN JUAN
RIO ARRIBA
TAOS
COLFAX
UNION
MORA
HARDING
LOS ALAMOS
McKINLEY
SANDOVAL
● Santa Fe
SANTA FE
SAN MIGUEL
25
BERNALILLO
40
QUAY
CIBOLA
● Albuquerque
GUADALUPE
VALENCIA
CURRY
TORRANCE
DE BACA
CATRON
SOCORRO
LINCOLN
ROOSEVELT
25
CHAVES
SIERRA
LEA
GRANT
OTERO
EDDY
DONA ANA
● Las Cruces
10
LUNA
HIDALGO

New Mexico and Its Counties

Oklahoma and Its Counties

Texas and Its Counties

NATIVE AMERICAN
PLACENAMES
OF THE SOUTHWEST

A

A'AI STO (AZ, Pima Co.) A Tohono O'odham (Papago) community; the Tohono O'odham name *a'ai sto* means 'on-both-sides white', describing the surrounding valley.

AASAYII Wash (AZ, Navajo Co.) \ah SAH yee\. Probably from Navajo *ásaayi'* 'in a bowl', from *ásaa'* 'bowl'. A related name is **Asayi** (NM, McKinley Co.).

ABIQUIU (NM, Rio Arriba Co.) \AH bi kyoo, ah bi KYOO\. The Spanish town was founded in the 1500s on the site of a Tewa village called *p'ešu(-mbu'u)* 'timber-end (town)', but the source of the name *Abiquiu* is unknown. The modern town was made famous as the longtime residence of the late artist Georgia O'Keeffe, whose house is now open to the public as a museum.

ABO (NM, Torrance Co.) \ah BOH\. From the extinct Tompiro language; said to have meant 'water bowl'. First visited by Spaniards in 1598. The archaeological site is part of the Salinas Pueblo Missions National Monument.

ACALA (TX, Hudspeth Co.) \uh KAL uh\. The name is from Acala cotton, a variety originating in Acala, Chiapas, Mexico. The Mexican placename is from Aztec *acallan* 'place of boats', from *acalli* 'boat'.

ACHI (AZ, Pima Co.) \AH chee\. A village on the Tohono O'odham (Papago) Reservation; the name *aji* means 'thin, narrow'. This is the traditional site of a special prayer stick ceremony.

ACOMA \AH koh muh, AK oh muh, AK uh muh\. From Spanish *Ácoma*, from Keresan *áak'úume* 'a person from Acoma', containing *áak'u* 'Acoma pueblo'. **Acoma Pueblo** (NM, Cibola Co.), with its adjacent visitor center and museum, is the ceremonial center of the **Acoma Indian Reservation**. Because of its location on top of a mesa, Acoma is sometimes called "Sky City." There is also an **Acoma** in Arizona (Maricopa Co.).

ACOMILLA (NM, Socorro Co.) \ah kuh MEE yuh\. The name of this historic site is a Spanish diminutive of *Ácoma*, the term for the pueblo.

ACOMITA (NM, Cibola Co.) \ah kuh MEE tuh\. An alternative Spanish diminutive of *Ácoma*. This village was settled by residents of nearby Acoma Pueblo around 1870, when danger of raids by Apaches and Navajos no longer existed.

ADAHCHIJIYAHI Canyon (AZ, Navajo Co.) From Navajo *adah ch'íjíyáhí* 'where someone walked off a cliff', containing *adah* 'downward'.

ADAIR (OK, Mayes Co.) \AY dar\. A Cherokee family name. Named after William Penn Adair, a Cherokee leader. There is also an **Adair County**, Oklahoma.

ADULTERY Dune (AZ, Apache Co.) Corresponds to Navajo *séí adiléhé* 'adultery sand', from *séí* 'sand' and *adilyé* 'adultery'. The dune was a secluded place visited by illicit lovers.

AGATHLA Peak (AZ, Navajo Co.) \uh GATH luh\. From Navajo *aghaałą́* 'animal hair', from *aghaa* 'hair of animal, wool' and *łą́* 'much'. A tradition holds that Indians killed an antelope herd near here and scraped the hair from the hides against rocks at the base of this volcanic peak.

AGAWAM (OK, Grady Co.) \A guh wahm\. A transfer name from Agawam, Massachusetts (Hampden Co.); from southern New England Algonquian, containing *-woonki* 'crooked'.

AGUA SAL Creek (AZ, Apache Co.) Corresponds to Navajo *tó dík'ǫ́ǫ́zh ch'íníłį́*, 'salty water flows out' (lit. 'water salt flows-out').

AGUICO (NM, Cibola Co.) \HAH wi koo\. This archeological site was once a Zuni village; it was encountered by Spanish explorers in 1539. The name was earlier written **HAWIKUH** and is now often written **Hawikku**.

AHAN OWUCH (AZ, Pima Co.) A Tohono O'odham (Papago) village; the name *a'an owij* is literally 'feather awl'.

AHE VONAM (AZ, Pima Co.) A Tohono O'odham (Papago) village. The Tohono O'odham name *a'ai wonamĭ* means 'hat on-both-ends', supposedly because one of the nearby Brownell Mountains looks like a hat when viewed from either side.

AHLOSO (OK, Pontotoc Co.) \uh LOS oh\. From the Chickasaw placename *haalooso*. A post office here was so named between 1904 and 1917.

AHOL SAH (AZ, Coconino Co.) Perhaps from Navajo *ahoodzą́* 'hole, cavity'.

AHPEATONE (OK, Cotton Co.) \uh PEE uh tone\. The name of a post office from 1907 to 1916; said to be from Kiowa *á:fîtâugàu* 'lance-wood', referring to the Kiowa leader called Wooden Lance in English.

AH-SHI-SLE-PAH Wash (NM, San Juan Co.) \ah shee SHLEP uh\. From Navajo *áshįįh łibá*, lit. 'salt-gray'. Part of the colorful landforms here are administered as a wilderness area by the Bureau of Land Management.

AHWATUKEE (AZ, Maricopa Co.) \ah wah TOO kee\. This settlement was built around a house called in Spanish *Casa de Sueños* 'house of dreams'; the name is said to represent a translation of that phrase into Crow, a Siouan language of Montana. The closest counterpart identifiable in Crow, however, is *awachúhka* 'flat land, prairie'.

AJO (AZ, Pima Co.) \AH ho\. Perhaps a Spanish spelling of the Tohono O'odham (Papago) word for a red ore used for painting one's body; it has nothing to do with Spanish *ajo* 'garlic'. The post office was established

4

in 1900 at Old Ajo, but when that town burned to the ground, the name was transferred to the nearby settlement of Cornelia. Ajo was the home of the first copper mine in the state, and visitors can now view the New Cornelia Open Pit Mine, nearly two miles wide.

AK CHIN (AZ, Pinal Co.) \ahk CHIN\. A village on the Maricopa Ak Chin Indian Reservation, south of Phoenix; the Pima name *akĭ ciñ* means 'arroyo mouth'.

AK CHUT VAYA (AZ, Pima Co.) \AHK tit VWAH ya, AHK chit vwahya\. On the Tohono O'odham (Papago) Reservation; from Tohono O'odham *akĭ ched wahia* 'arroyo well'.

AK KOMELIK (AZ, Pima Co.) \ahk KO muh lik\. On the Tohono O'odham (Papago) Reservation; the Tohono O'odham name *akĭkomalik* is literally 'arroyo flat'.

ALABAMA. This word, the name of a Muskogean-speaking tribe affiliated with the Muskogee, was also given to the state of Alabama. The name is probably from a Choctaw term meaning 'plant cutters'. The present-day Choctaw form is *albaamu,* from *albah* 'plant' and *amo* 'to clear'; the word *albah* may refer specifically to medicinal plants. Descendants of the Alabama tribe currently live on the **Alabama-Coushatta Reservation** near Livingston, Texas (Polk Co.), and in the **Alabama-Quassarte Tribal Town**, affiliated with the Creek Nation, at Henryetta, Oklahoma (Okmulgee Co.). There are streams named **Alabama Creek** in Texas (Trinity Co.) and Oklahoma (Okfuskee Co.).

ALCHESAY Canyon (AZ, Maricopa Co.) \al chi SAY\. In 1917 this site was given the name of a famous White Mountain Apache leader, *ałchísé,* meaning 'little one'.

ALI AK CHIN (AZ, Pima Co.) \ah lee ahk CHIN\. On the Tohono O'odham (Papago) Reservation; the Tohono O'odham name *ali akĭ ciñ* means 'little arroyo mouth'.

ALI CHUK (AZ, Pima Co.) \ah lee CHUK\. On the Tohono O'odham (Papago) Reservation; the Tohono O'odham name is *ali jeg* 'small hole'.

ALI CHUKSON (AZ, Pima Co.) \ah lee CHOOK son\. A village on the Tohono O'odham (Papago) reservation; the Tohono O'odham name *ali cukṣon* 'little black base' refers to the foot of a black lava hill. A Spanish name for the site is *Tucsoncito* 'little Tucson', reflecting the fact that the name **TUCSON** is derived from the Native term *cukṣon* 'black base'.

ALIKCHI (OK, McCurtain Co.) \ah LIK chee, ah LIK shee\. From Choctaw *alikchi* 'doctor,' a reference to the nearby sulfur springs. During the period 1864–1907, Alikchi was the seat of **NASHOBA County** in **APUKSHUNNUBBEE District** of the Choctaw Nation. A post office was there between 1888 and 1931.

ALI MOLINA (AZ, Pima Co.) \ah lee mo LEE nuh\. On the Tohono O'odham (Papago) Reservation; the term *ali mali:na* 'little Magdalena' refers to a town in Mexico named Magdalena.

ALI OIDAK (AZ, Pima Co.) \ah lee OI dahk\. A Tohono O'odham (Papago) community; in Tohono O'odham *ali oidag* means 'little field'.

ALI WUA Pass (AZ, Pima Co.) \ah lee WOO uh\. The Tohono O'odham (Papago) name *ali wua* means 'little pond'.

ALLUWE (OK, Nowata Co.) \AL oo way\. From Delaware *alëwi* 'more'. A post office had this name between 1883 and 1909.

AL TSE TOH (AZ, Apache Co.) \ahl SEE to\. Apparently from Navajo *áłtsé tó*, lit. 'first water'.

ALUMA, Lake (OK, Oklahoma Co.) \uh LOO muh\. This reservoir and municipality, adjoining Oklahoma City on the northeast, was originally a game preserve. The name is said to be a shortening of Choctaw *aaloma cholosa* 'peaceful retreat'.

ALZONA Park (AZ, Maricopa Co.) \al ZOH nah\. A government housing project built on the outskirts of Phoenix during World War II; the name was formed by combining the "Al" from the Alcoa Aluminum Company with the last part of the name *Arizona*.

AMABALA (OK, Okfuskee Co.) \am uh BA la\. A reversed spelling of the name *Alabama*. That state is named after a Muskogean-speaking Indian tribe associated with the Muskogee; the term may be from Choctaw *albaamu* 'plant cutters'. An Amabala post office existed from 1900 to 1907.

AMOLA Ridge (NM, Mora Co.) Perhaps based on a form of *amole*, the Spanish name of a plant, sometimes called soaproot, ultimately from Aztec *amolli*.

AMOLE Peak (AZ, Pima Co.) \uh MOH lee\. The Spanish name of a plant, sometimes called soaproot, ultimately from Aztec *amolli*. The name occurs elsewhere in **Cañada de Amole** (NM, Santa Fe Co.), in **Los Amoles** (NM, Doña Ana Co.), which used the Spanish plural form, and perhaps in **AMOLA Ridge** (NM, Mora Co.).

AMOXIUMQUA (NM, Sandoval Co.) \uh MAHK see uhm kwah\. The Spanish spelling reflects Jemez *amun-sho-kwa* 'ant hill'.

AMUSOVI Mesa (AZ, Navajo Co.) From Hopi *angwus'ovi* 'high raven place'; from *angwusi* 'raven', *ooveq* 'high', and *-vi* 'place'.

ANACACHO (TX, Kinney Co.) \an uh KAH choh\. Perhaps an Indian name, but the derivation is not clear. The **Anacacho Mountains** are nearby.

ANADARCHE Creek (OK, Love Co.) \an uh DAHR kee\. From the same origin as **ANADARKO**.

ANADARKO (OK, county seat of Caddo Co.) \an uh DAHR ko\. From Caddo \nadá:kuh\ 'bumblebee place', referring to the Caddo-speaking Bumblebee people. The post office was established here in 1873. There is also an **Anadarko** in Texas (Rusk Co.). A related name is **Anadarche Creek.**

ANAHUAC (TX, seat of Chambers Co.) \AN uh wahk\. This name, dating from 1870, was transferred from Mexico; the Aztecs applied the word *anahuac,* lit. 'near the water', to their territory in central Mexico. The Spanish or Aztec pronunciation is \ah NAH wahk\. The **Anahuac National Wildlife Refuge** in Texas is on Galveston Bay.

ANALCO (NM, Santa Fe Co.) \uh NAL koh\. This neighborhood of Santa Fe was settled around 1610 by Indians whom Spanish colonists brought from Tlaxcala, Mexico. They spoke Aztec, and it was they who brought the name *analco* 'on the other side of the water'. This is the oldest settled area in Santa Fe, apart from the area immediately around the central plaza.

ANASAZI Ruins (AZ, Apache Co.) \ah nuh SAH zee\. The term *Anasazi* designates pueblo dwellers who flourished from around A.D. 100 to 1300. The name was borrowed from Navajo *anaasází* 'ancestors of enemies', referring to the modern Pueblo peoples of the Southwest as enemies of the Navajos; from Navajo *ana'í* 'enemy' and *asází* 'ancestor'.

ANEGAM (AZ, Pima Co.) On the Tohono O'odham (Papago) Reservation; from Tohono O'odham *aanegam* 'desert broom plants'.

ANNONA (TX, Red River Co.) This is the botanical name for the fruit genus otherwise known as the sweetsop or cherimoya; it is from Spanish *anona,* borrowed from an Arawakan language of the Caribbean area.

APACHE \uh PACH ee\. This is a cover term for several Indian peoples speaking languages related to Navajo. The Spanish form *apache* was first recorded in New Mexico in 1598; it is derived not from any Apachean language but from Yavapai *'paacha* 'people'. At present, members of the **Western Apache** tribe live in Arizona on the **Fort Apache** and **San Carlos Indian reservations** and on the **Tonto Indian Reservation** near Payson (Gila Co.). The **Mescalero Apache** tribe lives on the **Mescalero Indian Reservation** in New Mexico (Otero Co.). Members of the **Lipan Apache** tribe also live on the Mescalero Reservation. The **Jicarilla** Apache tribe lives on the **Jicarilla Apache Indian Reservation** in northwestern New Mexico. The **Chiricahua Apache** tribe lives partly on the Mescalero Reservation in New Mexico and partly in the **Fort Sill Apache** community near Lawton, Oklahoma (Comanche Co.). The **Plains Apaches** (formerly known as Kiowa-Apaches) are represented by the **Apache Tribe of Oklahoma**, located at Anadarko, Oklahoma (Caddo Co.). The placename **Apache** occurs in many states; examples are the settlement of **Apache** in Oklahoma (Caddo Co.), **Apache County** (AZ), **Fort Apache** (AZ, Navajo Co.), **Apache Peak** and **Apache Lake** (AZ, Maricopa Co.), and **Apache Peak** (NM, Colfax Co.).

APACHERÍA (AZ, Gila Co.) \uh pach uh REE uh\. Spanish for 'Apache country'; see **APACHE**.

APONIVI (AZ, Navajo Co.) \uh POH nee vee\. From the Hopi placename *apòonivi,* of unclear meaning.

APPALACHIA (OK, Pawnee Co.) \ap uh LAY chee uh\. The exact location of this historic site, adjoining Keystone, is not on record. From *Apalachee,* the name of a Muskogean-speaking Indian tribe originally living in northwestern Florida; perhaps derived from Apalachee *abalahci* 'other side of the river'. The name of the **Appalachian Mountains,** in the eastern United States, is derived from the same source, as is, probably, the name of the Florida town **Appalachicola**.

APPALOOSA Ridge (AZ, Coconino Co.) \ap uh LOO suh, ah puh LOO sah\. Named after a breed of horse traditionally associated with the Nez Percé tribe of Idaho and the Palouse River of their region. The term is perhaps better associated, however, with the placename **Opelousas** (Louisiana, St. Landry Parish). *Palouse* is from Sahaptin *palú:s* 'what is standing up in the water'; *Opelousas* may be from Choctaw *api losa* 'black body'.

APUKSHUNNUBBEE District (OK) \ah puhk shuh NUH bee\. Named after a prominent Choctaw leader whose name included *abi* 'kill'. The area was part of the Choctaw Nation in the period 1864–1907; it corresponds to several present-day counties in the extreme southeastern corner of Oklahoma.

AQPI (AZ, Navajo Co.) This name of a spring may be from Hopi.

ARANSAS (TX) \uh RAN zus\. The name originally referred to the channel between Mustang and St. Joseph's islands; the Indian term *aranzu* (perhaps from the now extinct language Karankawa) was applied by a Spanish explorer in 1746. The name was later applied to **Aransas County,** to the **Aransas River** (Refugio and San Patricio Cos.), and to the towns of **Aransas Pass** (San Patricio Co.) and **Port Aransas** (Nueces Co.), all in Texas. **Aransas National Wildlife Refuge** is in adjacent Calhoun County.

ARAPAHO (OK, county seat of Custer Co.) \uh RAP uh hoh\. The name refers to a Plains Indian tribe of the Algonquian language family; at present one branch lives in Wyoming, and the other, near Concho, Oklahoma (Canadian Co.). The name of the tribe may have been borrowed by whites from Crow *aa-raxpé-ahu,* 'tattoo'. The headquarters of the **Cheyenne-Arapahoe Tribes of Oklahoma** is at Concho, Oklahoma (Canadian Co.).

ARAVAIPA (AZ, Graham Co.) \air uh VIE puh, ah ruh VIE puh\. The name was first given to **Aravaipa Creek,** probably from Tohono O'odham (Papago) *ali waippia* 'little wells' (plural of *wahia* 'well'). The post office dates from 1892.

ARBECA Creek (OK, Hughes Co.) \ahr BEE kuh\. From the Muskogee tribal town *apehkv,* originally located in Georgia. The variant spelling **Arbeka** designates a historic post office site (OK, Seminole Co.). There is also an Arbeka Church (OK, Okfuskee Co.).

ARIVACA (AZ, Pima Co.) \air uh VAK uh, air uh VAH kuh\. A Tohono O'odham (Papago) community; the Tohono O'odham name *ali wa:k* means 'little water'. The name was first recorded by Spaniards in 1773.

ARIZOLA (AZ, Pinal Co.) \air uh ZOH luh\. A combination of *Arizona* and *Ola,* the name of the founder's daughter.

ARIZONA (state) \air uh ZOH nuh\. The name of the state is from Spanish *Arizonac,* probably derived from Tohono O'odham (Papago) *ali ṣonag* 'having a little spring'. The term was recorded in Spanish in 1754, referring to a site in Mexico, just south of what is now the border. However, a derivation from Basque *arizonak* 'good oaks' is also possible, because Basque miners from Spain were early settlers in the area. There is also a place called **Arizona** in Walker County, Texas.

ARKANSAS River \AHR kun sah, AHR kun saw\. This stream flows through the states of Colorado, Kansas, Oklahoma, and Arkansas, to which it gives its name. Most of the state of Oklahoma is within the drainage of the Arkansas River. The placename is from French *Acansas,* a name applied to the Quapaw, a southern Siouan tribe who lived on this river. The French term was derived from Algonquian *akansa,* referring to the southern Siouan peoples, and this in turn was derived from the Siouan stem *kką́ze,* which also entered English as *Kansas.* There is an **Arkansas City** in Starr County, Texas, and an **Arkansas** (pronounced \ahr KAN zus\) **City** in Cowley County, Kansas.

ARKOMA (OK, Le Flore Co.) \ahr KOH muh\. A blend of *Arkansas* and *Oklahoma*; the post office was established in 1914.

ARTUSEE (OK, McIntosh Co.) A Creek tribal town in the present-day town of Eufaula; from Muskogee *atvse.*

ASAI STO (AZ, Pima Co.) Perhaps a variant writing of **A'AI STO**, from Tohono O'odham (Papago) *a'ai sto* 'on-both-sides white'.

ASAYI (NM, McKinley Co.) \uh SAH yee\. From Navajo *ásaayi'* 'inside the bowl', from *ásaa'* 'bowl'. Related names are **Asayi Lake** (NM, McKinley Co.), **Aasayii** (AZ, Navajo Co.), and **Aasayii Wash** (AZ, Pima Co.).

ASAYVA (AZ, Navajo Co.) \uh SIE vuh\. From Hopi *asayva* 'sparrow spring', from *asay* 'sparrow'.

ASSAN DECHALI Spring (AZ, Navajo Co.) \uh san duh CHAH lee\. Possibly represents Navajo *asdzáán dijoolí,* lit. 'woman who-is-round'.

ASTIALKWA (NM, Sandoval Co.) A former Towa pueblo site; from Jemez (Towa) *aa-tyole-kwa* 'place for bringing down a grinding-stone'.

ATA DEEZA (AZ, Apache Co.) Navajo *atah deez'á* 'among ridges or bluffs', from *atah* 'among' and *deez'á* '(a long ridge) to lie'.

ATOKA (OK, county seat of Atoka Co.) \uh TOH kuh\. Said to commemorate an Indian leader called Captain Atoka; perhaps from Choctaw *hitoka* 'ball-ground', that is, a field on which a Native game similar to field hockey was played. Between 1864 and 1907, the county was part of **PUSHMATAHA** District in the Choctaw Nation. **Atoka Reservoir** and the **Atoka Wildlife Management Area** are nearby. **Atoka** also occurs as a transfer name in New Mexico (Eddy Co.).

ATOKO Point (AZ, Coconino Co.) \uh TOH koh\. In Grand Canyon National Park; from Hopi *atoko* 'crane or other long-legged wading bird'.

ATOLE Pond (NM, Rio Arriba Co.) \uh TOH lee\. From Spanish *atole* 'a beverage made from cornmeal', from Aztec *atolli.*

ATRISCO (NM, Bernalillo Co.) \uh TRIS koh\. This name of an Albuquerque neighborhood was imported by Aztec-speaking settlers brought by Spaniards from Tlaxcala state, Mexico, around 1703. It is probably from Aztec *atlixco* 'on the surface of the water'.

ATSADAHSIDAHI (AZ, Apache Co.) \uh tsah dah si dah HEE\. From Navajo *atsá dah sidáhí* 'where the eagles sit on high', from *atsá* 'eagle', *dah* 'up', and *niidáh* 'to sit in a row'.

ATSEE NITSAA (NM, McKinley Co.) \uh TSAY nit sah\. Apparently from Navajo *atsee' nitsaa,* 'big tail' (lit. 'tail-big').

ATSINNA (NM, Cibola Co.) \AH tsin uh\. The name of this ancestral Zuni village (now in ruins) in El Morro National Monument means 'writing on the rocks', corresponding to the English name *Inscription Rock.* On the soft sandstone here one can see petroglyphs carved by prehistoric Indians, followed by the 1605 inscription of Juan de Oñate, the first Spanish visitor, and signatures by hundreds of later travelers.

ATTOYAC (TX, Nacogdoches Co.) \AT uh yak\. Perhaps imported from Mexico, where several places are named **Atoyac**, from Aztec 'at the river' (*atoyatl* 'river'). **Attoyac BAYOU** is near the Texas town of Attoyac.

AWATOBI (AZ, Navajo Co.) \uh WAH tuh bee\. The name of these ruins, also spelled **Awatovi**, is from Hopi *awat'ovi* 'high place of the bow', because the Bow Clan is said to have lived there. The community is said to have been destroyed in 1700 by Hopis from other villages, in a conflict over conversion to Christianity. Related names are **Awatovi Spring** (AZ, Navajo Co.) and **Awatubi Creek** (AZ, Coconino Co.).

AWATOVI Spring (AZ, Navajo Co.). From Hopi *awat'ovi* 'high place of the bow'; see **AWATOBI**.

AWATUBI CREEK (AZ, Coconino Co.). From Hopi *awat'ovi* 'high place of the bow'; see **AWATOBI**.

AZANSOSI Mesa (AZ, Navajo Co.) \uh zahn soh SEE\. Named for Louise Wetherill, the wife of a trader in the area. Her Navajo name was *asdzą́ą́-ts'ósí* 'slim woman' (lit. 'woman-slim').

AZTEC (NM, San Juan Co.) \AZ tek\. The English name of the Aztec people, who dominated much of central Mexico at the time of the Spanish conquest. It comes from Spanish *Azteca,* which in turn is from Aztec *Aztecah*, meaning 'people of *Aztlan'*, a legendary homeland. English-speaking settlers applied the name in the 1870s to the nearby ancestral Pueblo ruins now called **Aztec Ruins National Monument**; it was mistakenly believed that the ruins had been built by Indians related to the Aztecs of Mexico. **Aztec Peak** in Arizona (Gila Co.) was so named in the same mistaken belief that Aztecs had once lived there.

B

BABOCOMARI River (AZ, Cochise Co.) \bah boh KOH muh ree, bah boh kuh MAH ree\. This name was first given to a Spanish land grant in 1832, in the spelling *Babacomari.* It is from Tohono O'odham (Papago) *waw komali,* lit. 'bedrock flat', containing *waw* 'bedrock'.

BABOQUIVARI Peak (AZ, Pima Co.) \bah boh KEE vuh ree, bah boh kwi VARE ee\. A Spanish spelling of Tohono O'odham (Papago) *waw giwulĭk* 'bedrock constricted in the middle', containing *waw* 'bedrock'.

BABY Rocks (AZ, Navajo Co.). So called because the sandstone here appears to be composed of many small rocks, rather than continuous layers. The name is a translation of Navajo *tsé awé'é* 'rock babies'.

BABY PEE Trail (AZ, Apache Co.) A translation of Navajo *awéé haazhilzhí* 'where a baby urinated on the way up'.

BAC, San Xavier del (AZ, Pima Co.). See **SAN XAVIER DEL BAC, Mission.**

BACAVI (AZ, Navajo Co.) \BAH kuh bee, BAH kuh vee\. A Hopi village; from *paaqavi* 'place of reeds', containing *paaqa* 'reed'. Variants are **Bacobi** and **Bakabi.**

BACOBI (AZ, Navajo Co.) \BAH kuh bee, BAH kuh vee\. See **BACABI.**

BAHA Spring (AZ, Navajo Co.) This perhaps represents Navajo *bąąhí* 'place where there is an edge', from *bąąh* 'edge'.

BA HA ZOHNNIE BETOH (AZ, Navajo Co.) \bah huh ZHOH nee bi TOH\. From Navajo *bąąh hózhóní bito'* 'alongside a pleasant spring', containing *hózhóní* 'it is pleasant' and *bito'* 'its water, its spring'.

BAKABI (AZ, Navajo Co.) See **BACAVI.**

BALAKAI Mesa (AZ, Apache Co.) \bah luh KIE\. From Navajo *baalók'aa'í* 'reeds alongside it', from *baa* 'alongside it' and *lók'aa'* 'reeds'. Also written **Balokai** and **Balukai.**

BAPCHULE (AZ, Pinal Co.) \bap CHOO luh\. On the Gila River Indian Reservation; perhaps from a Pima term meaning 'hooked nose'.

BAYOU \BAH yoo, BAH yoh\. This English term for a sluggish stream is borrowed from French, which took it from an early Choctaw form, *bayuk* 'creek, river'. **Bogue,** a related form, is from modern Choctaw *book* 'creek'. In specific placenames of eastern Texas and Oklahoma (as in neighboring Louisiana), *Bayou* generally occurs before a modifying noun, on the French pattern; examples are **Bayou Carrizo** (TX, Nacogdoches Co.), with Spanish *carrizo* 'cane grass', and **Bayou Manard** (OK, Muskogee Co.), with the French family name *Manard.* But in some cases the word *Bayou* comes second, as in **ATTOYAC Bayou** (TX, Nacogdoches Co.) and **CHOCOLATE Bayou** (TX, Brazoria Co.).

BECLABITO (AZ, San Juan Co.) \bi KLAH bi toh\. A variant of **BITLABITO.**

BEGAY Well (AZ, Apache Co.) \bi GAY\. The word Begay, a common surname among English-speaking Navajos, is from *bighe', biye'* 'his son'. Originally, Navajos did not use surnames, but when Anglo administrators asked their names, they sometimes got answers like *Pedro bighe'* 'son of Pedro'. The administrators interpreted *bighe'* as a surname, writing it as *Begay*. For a parallel case, see **BENALLY.**

BEKIHATSO Lake (AZ, Apache Co.) From Navajo *be'ekid hatsoh,* lit. 'lake big-area'.

BENALLY (AZ, Coconino Co.) \bi NAH lee, bi NAL ee\. Like **BEGAY**, this is a common surname among English-speaking Navajos; it is from *binálí* 'his grandson'. Originally, Navajos did not use surnames, but when Anglo administrators asked their names, they sometimes got answers like *Pedro binálí* 'grandson of Pedro'. The administrators interpreted *binálí* as a surname, writing it as Benally.

BESHBAGOWAH Ruins (AZ, Gila Co.) \besh bah GOH wuh\. Although these are the ruins of an ancestral Pueblo settlement, the name is from Western Apache *bésh baaghowqh* 'house for metal', with *bésh* meaning 'flint, metal', which probably refers to the ore smelter at Globe, Arizona. The archaeological site has 300 rooms, inhabited during the period 1225–1400.

BESHBITO Wash (AZ, Navajo Co.) \BESH bi toh\. From Navajo *béésh bito'* 'spring of flint or metal', containing *béésh* 'flint, metal'.

BETATAKIN (AZ, Navajo Co.) \buh TAH tuh kin, buh TAT uh kin\. This is the ruin of an ancestral Pueblo settlement, but it bears the Navajo name *bitát'ah kin,* lit. 'on-rock-ledge house'. Hopis say that the Navajo term is borrowed from Hopi *pitataki* 'houses clinging'. The site is now part of **Navajo National Monument.**

BETAY (AZ, Coconino Co.) \buh TIE, buh TAY\. The name of this summit is perhaps from Navajo *bidáá'í* 'its edge, its ledge'.

BETONNIE TSOSIE Wash (NM, San Juan Co.) \buh TOH nee TSOH see\. Named after a Navajo man called *baadaaní ts'ósí* 'her slim son-in-law' (*ts'ósí* means 'slim'), or, in English, Slim Bitani. See **BITANI TSOSIE Wash**, which apparently uses an alternate spelling.

BETONY Butte (AZ, Navajo Co.) \buh TOH nee\. Probably from the surname *Bitani,* as in the previous entry. This family name is written in English as *Betonnie, Bitani, Betony,* and so forth.

BI KEESH Wash (AZ, Apache Co.) \bi KEESH\. Perhaps from Navajo *bik'iish* 'its alder tree', containing *bi-* 'his, her, its' and *k'iish* 'alder tree'.

BIBYAK (AZ, Pima Co.) \BIB ee yahk\. This Tohono O'odham (Papago) name may be from *bibbi* 'plates of food' and *akǐ* 'arroyo'.

BIDAAII TO HAALI (AZ, Coconino Co.) \bi DAH ee toh HAH lee\. This spring has the Navajo name *bidáá'í tó háálį* 'ledge where water flows up and out', containing *háálį* 'it flows up and out'.

BIDAHOCHI (AZ, Navajo Co.) \BID uh hoh chee, bid uh HO chee\. From Navajo *bidahóóchii'* 'a red area coming down, extending downward'. Also written **Bidahochee.** Named for a slide of red rock behind the trading post founded there in 1882.

BIHILINIE Canyon (AZ, Apache Co.) \bi huh LEE nee\. Perhaps from Navajo *bįįh háálíní* 'deer spring', containing *bįįh* 'deer'.

BIJAADIBAE (AZ, Apache Co.) \bi jah di BIE\. Perhaps from Navajo *bijáádíbái* 'gray legs', containing *bijáad* 'leg'.

BIKO HODO KLIZH (AZ, Coconino Co.) \bi koh ho DOH klizh\. This valley takes its name from Navajo *bikooh hodootl'izh*, lit. 'canyon blue-space', which combines *bikooh* 'canyon' and *dootl'izh* 'blue'.

BINNE ETTENI Canyon (AZ, Coconino Co.) \bi NEE ET uh nee\. Probably from a Navajo personal name such as *binii'ádinii*, lit. 'his-face does-not-exist', containing *binii* 'his face'.

BIS II AH Wash (AZ, Apache Co.) \bis EE ah\. From Navajo *bis íí'á*, lit. 'adobe it-stands-up', containing *bis* 'adobe'.

BISTI (NM, San Juan Co.) This eroded landscape, or "badlands," takes its name from Navajo *bistahí* 'among the adobe formations', containing *bis* 'adobe'.

BITANI TSOSIE Wash (NM, San Juan Co.) \bi TAH nee SOH see\. Named after a Navajo livestock owner called, in English, Slim Bitani. The Navajo name of Bitani Tsosie Wash is *k'ai' naashchii' bikooh,* lit. 'willow meandering-red-line its-canyon', containing *k'ai'* 'willow' and *kooh* 'canyon'. See **BETONNIE TSOSIE Wash**, which apparently uses an alternate spelling.

BITLABITO (AZ, Apache Co.) \bi KLAH bi toh\. From Navajo *bitlááh bito'* 'water underneath'; containing *bitlááh* 'underneath'. Also written **Beclabito.**

BITSIHUITOS Butte (AZ, Apache Co.) \bit see WEE tos\. From Navajo *bitsį́į́h hwiits'os* 'a hill that tapers at its base', containing *bitsį́į́h* 'its base'.

BIXHOMA Lake (OK, Wagoner Co.) Probably from the family name *Bixby* plus the latter part of the word *Oklahoma*.

BLACK KETTLE National Grassland (OK, Roger Mills Co.) Named for Black Kettle, the Cheyenne leader who was defeated there in 1868 by Lt. Col. George Custer, on the Washita River, in what the U.S. government calls the Battle of the Washita and the Cheyenne and Arapaho Tribes call the Black Kettle Massacre. The **Black Kettle Museum**, in Cheyenne, Oklahoma, documents the tribal history.

BODOWAY Mesa (AZ, Coconino Co.) \BOD uh way\. Named for a nineteenth-century leader of the Southern Paiute people. His Native name was *potokwai* 'they travel away', from *poto* '(pl.) to travel'.

BOGATA (TX, Red River Co.) \buh GO tuh\. It was intended to name this town after Bogotá, Colombia (from the Chibcha language of that area). The current pronunciation reflects the vowel qualities of the intended spelling.

BOGUE Springs (TX, Cass Co.) From Choctaw *book* 'creek'. An earlier Choctaw form, *bayuk,* is the source of English **BAYOU.**

BOKCHITO (OK, Bryan Co.) \BOHK chee toh, bohk CHEE toh\. In the historic Choctaw Nation; from Choctaw *book chito,* lit. 'stream big'. See **BOGUE.**

BOKHOMA (OK, McCurtain Co.) \BOHK hoh muh, bohk HOH muh\. In the historic Choctaw Nation; from Choctaw *book homma,* 'red stream' (lit. 'stream red'). A post office existed here between 1904 and 1936.

BOKOSHE (OK, Le Flore Co.) \boh KOH shee\. In the historic Choctaw Nation; from Choctaw *book ushi* 'little stream' (lit. 'stream little'). The post office was established in 1886. **Bokoshe Mountain** is nearby.

BOKTUKLO (OK, McCurtain Co.) \bohk TOOK luh\. A county in **APUKSHUNNUBEE District** of the original Choctaw Nation; from Choctaw *book tuklo,* 'two streams' (lit. 'stream-two'). A post office existed here between 1908 and 1921. **Boktuklo Creek** flows into the Glover River in the same county.

BOWLEGS (OK, Seminole Co.) In the historic Seminole Nation; named for Billy Bowlegs, a Seminole leader. A post office was established here in 1927.

BOYSAG Point (AZ, Mohave Co.) \BOI sag\. The name of this point on the North Rim of the Grand Canyon is from Southern Paiute *païsaka* 'bridge'. It was originally so named because it could be reached only by a small, man-made bridge.

BRAZIL Creek (OK, Le Flore Co.) Flows into the Poteau River; named after the South American nation, whose name is of Indian origin.

BUNCHED FEATHERS (AZ, Apache Co.) In Canyon de Chelly National Monument. Corresponds to Navajo *t'á' nááséél* 'a crowd of people is moving back with feathers', containing *t'á'* 'feather'.

BURNT CORN Wash (AZ, Navajo Co.) Corresponds to Navajo *naadą́ą́ diilid,* lit. 'corn is-burnt', containing *naadą́ą́* 'corn'.

BURNT WATER (AZ, Apache Co.) Corresponds to Navajo *tó díílidí* 'water that is burnt', because a burnt ramada fell into a well here.

BURRO, Mount (AZ, Coconino Co.) Probably named for a Havasupai family with the surname *Burro.*

BUSHYHEAD (OK, Rogers Co.) In the historic Cherokee Nation; named for Dennis Bushyhead, Cherokee principal chief in 1879–87.

BYLAS (AZ, Graham Co.) \BIE lus\. This community on the San Carlos Apache Reservation has the name of a nineteenth-century Apache leader. The Native origin of the name has not been traced.

C

CABABI Mine (AZ, Pima Co.) \kah BAH bee, kuh BAH bee\. From Tohono O'odham (Papago), meaning 'badger well', containing *ka:w* 'badger' and *wahia* 'well'.

CADDO \KAD oh\. An Indian group that formerly occupied northeastern Texas, southwestern Arkansas, and the Red River Valley of Louisiana, southern Arkansas, and southeastern Oklahoma; the Caddos now live in Oklahoma. Their language belongs to a family that is named, after them, **Caddoan.** The term is from the first two syllables of *kaduhdá:ču'*, the name of one of the divisions of the Caddos; the last two syllables mean 'sharp'. In Oklahoma, the **Caddo Tribe** has its headquarters at Binger in **Caddo County**, and there is a town called **Caddo** in Bryan County. Texas has towns called **Caddo** (Wilson Co.) and **Caddo Mills** (Hunt Co.), and **Caddo Creek** flows into the **NECHES River** (Anderson Co.). **Caddo Lake** (Harrison Co.) is the largest natural lake in Texas, probably formed by the great New Madrid earthquake of 1811; it extends into Louisiana. Also in Texas are **Caddoan Mounds State Historic Site** (Cherokee Co.) and the **Caddo National Grassland** (Fannin Co.). There is also a place called **Caddo** in Arizona (Navajo Co.).

CALUMET (OK, Canadian Co.) \KAL yoo met, KAL yuh met, kal ya MET\. The word refers to the ceremonial pipe for smoking tobacco used by Indians of the central states. Although sometimes thought to be an Indian word, it is actually from a French dialect term, *calumet* 'pipe', derived from Latin *calamus* 'reed'.

CALVA (AZ, Graham Co.) \KAL vuh\. Named for a San Carlos Apache leader; perhaps short for the Spanish name *Calvario.*

CAMOTE (AZ, Pima Co.) \kuh MOH tee\. From Mexican Spanish *camote* 'sweet potato', from Aztec *camohtli.*

CANADIAN. This term, referring to a person from Canada, is common in U.S. placenames and in many cases may commemorate Canadian explorers and trappers; an example is the more northern of the two streams called the **Canadian River** in northern Colorado (Jackson Co.). However, the name of the **Canadian River** of southern Colorado, northern Texas, and Oklahoma is probably from Spanish *Río Canadiano*, a popular etymology from Caddo *káyántinu'*, used by the Indians to refer not to the Canadian River but to the nearby Red River. The **Canadian District** of the Cherokee Nation, between 1866 and 1907, was named for the river, which formed one of its boundaries. Oklahoma today has a town of **Canadian** (Pittsburgh Co.) and a **Canadian County** (with seat at El Reno), both also named for the river.

CANOE Mountain (OK, Cherokee Co.) \kuh NOO\. This word for a portable boat like those made by Native Americans was borrowed from French *canot,* from Spanish *canoa,* from an Arawakan language of the Caribbean.

CARANCAHUA (TX, Calhoun Co.) Named for the Indian tribe whose name is also written *Karankawa,* who once lived in northwestern Mexico and adjacent Texas; **Carancahua Creek** and **Carancahua Bay** are nearby. **Lake Carancahua** is in Galveston County.

CARNUEL (NM, Bernalillo Co.) \karn WEL\. A Spanish version of a Tiwa village name, *Carna-aye,* said to mean 'badger place'.

CASSADORE Mountain (AZ, Gila Co.) \KAS uh dor\. Named after a nineteenth-century Apache leader. The word may represent Spanish *cazador* 'hunter', from *cazar* 'to hunt'.

CATALE (OK, Rogers Co.) In the historic Cherokee Nation; said to mean 'in the valley'.

CATOOSA (OK, Rogers Co.) \kuh TOO suh\. In the historic Cherokee Nation; a transfer name from Catoosa County, Georgia, derived from Cherokee *gadusi* 'hill'.

CAYOTE (TX, Bosque Co.) \KAY oht\. From the same source as *coyote,* the name of an animal, from Mexican Spanish *coyote,* from Aztec *coyotl.* The word is written **COYOTE** in a large number of other placenames. In English the animal name is pronounced \KAH yoht\ or \kah YOH tee\ or \koy YOH tee\. Another spelling is **Kyote** (TX, Atascosa Co.).

CAYUGA (TX, Anderson Co.) \kah YOO guh\. A transfer name from Cayuga County, New York. From the Cayugas' name for themselves, *kayohkhó:nǫ',* or, less commonly, *kayokhwęhó:nǫ',* of unclear derivation. The headquarters of the **Seneca-Cayuga Tribe of Oklahoma** is in Miami, Oklahoma (Ottawa Co.). There are towns called **Cayuga** in Texas (Anderson Co.) and Oklahoma (Delaware Co.).

CEBOLLETA (NM, Cibola Co.) \seb oh YET uh\. Spanish for 'wild onion', a diminutive of *cebolla* 'onion'. The name corresponds to Navajo *tł'ohchin* 'onion', from *tł'oh* 'grass', perhaps related to the word *halchin* 'to have an odor'. The spelling **Seboyeta** is also used. **Cebolletita** (NM, Cibola Co.) means 'little wild-onion', so named because it was settled by people from Cebolleta.

CENTRAHOMA (OK, Coal Co.) \sen tra HOH muh\. Coined from the phrase *central Oklahoma.* The name **OKLAHOMA** itself was coined from Choctaw, meaning 'Indians' (lit. 'people-red').

CERBAT (AZ, Mohave Co.) Perhaps from the Maricopa (Yuman) word for 'bighorn sheep, mountain sheep'; the form is *svaat.* This and other words for mountain sheep in several southwestern Indian languages are borrowed from Spanish *chivato* 'he-goat', from *chivo* 'kid, young goat'.

18

CHACO Canyon (NM, San Juan Co.) \CHAH koh\. From Navajo *tsékooh* 'rock canyon', from *tsé* 'rock' and *kooh* 'canyon'.

CHALCHIHUITL, Mount (NM, Santa Fe Co.) \CHAHL chi wee tul\. From Aztec *chalchihuitl* 'precious green stone'.

CHAMA (NM, Rio Arriba Co.) \CHAH muh\. A shortening of Tewa *tsąmą'ǫŋwįkeyi* 'wrestling pueblo-ruin', from *tsąmą* 'to wrestle' and *ǫŋwįkeyi* 'pueblo ruin' (*ǫŋwį* 'pueblo', *keyi* 'old'). Nearby is **Chamita** (NM, Rio Arriba Co.) \chah MEE tuh\, the Spanish diminutive, meaning 'little Chama'.

CHAPOTE, EL, Creek (TX, Starr Co.) From a Mexican Spanish term referring to the *zapote prieto* ('dark zapote'), or Texas persimmon (*Diospyros texana*), a tree bearing an edible fruit. The word is derived from Nahuatl *zapotl* 'a fruit, the zapote'.

CHATTANOOGA (OK, Comanche Co.) \chat uh NOO guh\. A transfer name from Tennessee (Hamilton Co.). The name is from Muskogee *cvto-nokwv* 'rock-neck', probably borrowed first into Cherokee and then into English.

CHAT VAYA (AZ, Pima Co.) \chat VAH yuh\. On the Tohono O'odham Indian Reservation; probably containing Tohono O'odham *waya* 'well'.

CHECOTAH (OK, McIntosh Co.) \shi KOH tuh\. In the historic Creek Nation; named for Samuel Checote, a Muskogee leader. The Muskogee form of the placename is *cokotv*. Possibly related is **CHICOTA** (TX, Lamar Co.).

CHEDISKI Peak (AZ, Navajo Co.) \CHED i skie\. Probably from Western Apache *tsee deesgai* 'white stones extend horizontally', containing *tsee* 'stone' and *sgai* 'white'.

CHEE Well (NM, McKinley Co.) \chee\. Probably represents a Navajo given name, *chii,* short for *łichíí* 'red'. **Chee Dodge School** (NM, McKinley Co.) was named for a Navajo leader whose Native name was *hastiin adiits'a'ii'* 'the man who hears or understands'.

CHEECHILGEETHO (NM, McKinley Co.) \chee chil GEE toh\. Probably from Navajo *chéch'ilgitó* 'spring at the oak trees', from *chéch'il* 'oak tree' (*tsé, ché* 'rock' and *ch'il* 'plant') and *tó* 'water, spring'. Another current name for this place is **CHI CHIL TAH**.

CHENANGO (TX, Brazoria Co.) A transfer name from New York state, said to be from Onondaga *ochenang* 'bull thistles'.

CHEROKEE \CHER uh kee**,** CHAR uh kee\. An Indian group of the southeastern United States, of the Iroquoian linguistic family. At the time of first contact with whites, the Cherokees lived in the southern Appalachians. In the early nineteenth century, many of them were forced to resettle in what is now the **Cherokee Nation** of Oklahoma, although some remain on a reservation in North Carolina. The Native term for the group is *tsalagi*. As a placename, **Cherokee** is widespread in the United

States. In Oklahoma, the Cherokee Nation has its historic capital at Tahlequah, **Cherokee County.** In the same county is **Cherokee Wildlife Management Area. Cherokee State Park** is in Mayes County. **Cherokee Courthouse**, near Gore (Sequoyah Co.), is a historic building that once housed the government of the Cherokee Nation. There is also a town in Oklahoma called **Cherokee** (Alfalfa Co.). Texas, too, has a **Cherokee County** (with its county seat at Rusk) and a town called **Cherokee** (San Saba Co.).

CHETTRO KETTLE (NM, San Juan Co.) \chet roh KET ul\. This Navajo name is *tsé bidádi'ní'ání* 'rock that is plugged or sealed up', containing *tsé* 'rock', *bidádi'* 'blocking it', and *ní'á* 'solid objects lie in a row or line'. Also spelled **Chetro Ketl.**

CHEYARHA (OK, Seminole Co.) Name of a former Seminole tribal town once located in the present-day city of Seminole; from Muskogee *cēyahv.*

CHEYAVA Falls (AZ, Coconino Co.) \chay YAH vuh\. From Hopi *tsiyakva,* from *tsiyak-* 'to crack' and *-va* 'water'.

CHEYENNE \shie EN, shie AN\. An Indian people of the Algonquian linguistic family, living in the Great Plains. At present there are Cheyenne communities in Wyoming and Oklahoma. The English name comes from French *cheyenne,* originally recorded in 1679 as *Chaiena,* from Dakota (Siouan) *šahíyena.* This is the diminutive of *šahíya,* a Dakota name for the Cree people. The term is widespread as a placename. In Oklahoma, the town of **Cheyenne** is the county seat of Roger Mills County, and the headquarters of the **Cheyenne-Arapahoe Tribes of Oklahoma** is at Concho (Canadian Co.). Oklahoma also has places called **Cheyenne Butte** (Blaine Co.) and **Cheyenne Valley** (Major Co.).

CHIAWULI TAK (AZ, Pima Co.) On the Tohono O'odham (Papago) Indian Reservation; probably derived from Tohono O'odham *jiawul* 'devil'.

CHI CHIL TAH (NM, McKinley Co.) \CHEE chil tah\. From Navajo *chéch'il-tah,* lit. 'oaks-among'; the word *chéch'il* 'oak' is from *ché, tsé* 'rock' and *ch'il* 'plant'. This name refers to the same place as the related **CHEECHILGEETHO.**

CHICKASAW \CHIK uh sah, CHIK uh saw\. A people of the Muskogean language family, originally living in northern Mississippi and western Tennessee; the **Chickasaw Nation** is now in Oklahoma. The term comes from the Chickasaws' name for themselves, *chikashsha.* The Chickasaw Nation has its historic capital at **TISHOMINGO** (Johnston Co.); the **Chickasaw Council House Museum** there contains the log structure that once served as the seat of government for the Chickasaw Nation. The **Chickasaw National Recreation Area** and the **Chickasaw Wildlife**

Management Area are near Sulphur (Murray Co.). Also in Oklahoma are **Chickasaw Creek** (Atoka Co.) and **Chickasaw Lake** (Carter Co.).

CHICKASHA (OK, Grady Co.) \CHIK uh shay\. In the historic Chickasaw Nation; a variant of **CHICKASAW**. **Lake Chickasha** is near Gracemont (Caddo Co.). Present-day Chickasha was called **WACO** from 1890 to 1891, Pensee from 1891 to 1892, and then Chickasha, after the Chickasaw Tribe.

CHICO \CHEE koh\. This Spanish word means 'small' and can also be a nickname for someone called Francisco. In the Southwest it is used as the name of a native bush, in English called black greasewood; this is probably the source of the placename **Chico** in New Mexico (Colfax Co.). An Indian origin for the name of the bush is also possible. A Spanish adjective derived from the name of the plant appears in **CHICOSA Lake** (NM, Harding Co.).

CHICOMA Mountain (NM, Rio Arriba Co.) \chi KOH muh\. Perhaps from a Tewa word meaning 'flint, obsidian'.

CHICORICA Creek (NM, Colfax Co.) \chik uh REE kuh\. Perhaps from a Plains Apache term for 'turkey', equivalent to Navajo *tsídii łíkizh* 'spotted bird' (lit. 'bird-spotted'). This may have become Spanish *chiti-riqui,* changed by assimilation to *chiqui-riqui* and then reinterpreted through folk etymology to *chico rico* 'rich child'. English mispronunciation then resulted in *Chicorica.* Alternatively, *Chicorica* may be an adaptation of the name of the **JICARILLA** Apaches, from Spanish *jicarilla* 'little cup', diminutive of *jícara* 'gourd cup', from Aztec *xicalli.* The term *Chicorica* has itself been adapted into English as **SUGARITE** \SHOOG uh reet\.

CHICOSA Lake (NM, Harding Co.) \chi KOH zuh\. The Spanish adjective *chicosa* is derived from *chico* 'black greasewood', perhaps an Indian name in origin. The masculine form of the adjective, **CHICOSO** \chi KOH zoh\, also occurs as a placename in New Mexico (Colfax Co.).

CHICOSO (NM, Colfax Co.) \chi KOH zoh\. The Spanish adjective *chicoso* is derived from *chico* 'black greasewood', perhaps an Indian name in origin. The feminine form of the adjective, **Chicosa** \chi KOH zuh\, also occurs as a placename in New Mexico (Harding Co.).

CHICOTA (TX, Lamar Co.) \shi KOH tuh\ Possibly a variant of **CHECOTAH** (OK, McIntosh Co.). This is perhaps from a Muskogee placename, *cokotv.* Another possible origin is American Spanish *chicote* 'whip'.

CHIGLEY (OK, Murray Co.) \CHIG lee\. In the historic Chickasaw Nation; named for Nelson Chigley, a prominent Chickasaw.

CHIHUAHUA (TX, Hidalgo Co.) Named for a state in northern Mexico; the name is perhaps derived from Tarahumara *chihua-huasa* 'bag in the cornfield', from *chihua* 'bag' and *huasa* 'cornfield'. The **Chihuahuan**

Desert (TX, Hudspeth Co.) covers much of the state of Chihuahua and extends into western Texas.

CHIKASKIA River (OK, Kay Co.) \chi KAS kee uh, shi KAS kee uh\. This name has been said to be Osage for 'white spotted deer', but that origin cannot be confirmed. It may be an invented name, incorporating elements of two Illinois names, *Chicago* and *Kaskaskia*.

CHILEAN Mill (AZ, Yavapai Co.) \CHIL ee un\. From *Chile,* the South American country, a name probably derived from a local Indian language. A large number of miners came from Chile to California and Arizona during the early mining days.

CHILICOTAL Mountain (TX, Brewster Co.) \chil i koh TAHL\. Mexican Spanish, 'where the *chilicote* grows', from Aztec *chilicayotl* 'a squash-like plant'.

CHILICOTE Canyon (TX, Presidio Co.) \chil i KOH tee\. A variant of *chilicayote,* from Aztec *chilicayotl* 'a squash-like plant'.

CHILILI (NM, Bernalillo Co.) \chi li LEE\. Perhaps from a Tiwa word meaning 'sound of water barely trickling'.

CHILLI Creek (OK, Latimer Co.) \CHIL ee\. Perhaps from *Chile,* the South American country, or from Mexican Spanish *chile* 'chili pepper', from Aztec *chilli.*

CHILLICOTHE (TX, Hardeman Co.) \chil i KAW thee, chil i KOTH ee\. A transfer name from Ross County, Ohio. From Miami (or Illinois) *čalaka:θe* 'member of the *čalaka* subgroup', referring to a division of the Shawnee people (also Algonquian speakers, like the Miami).

CHILOCCO (OK, Kay Co.) \shi LOK oh\. From Muskogee *cerakko* 'horse', lit. 'big deer', from *eco* 'deer' and *-rakko* 'big'. **Chilocco Creek** is nearby, flowing into the Arkansas River. The appearance of a Muskogee (Creek) placename in a part of the state that is home to other tribes is perhaps explained by the fact that a large Indian boarding school, Chilocco Indian School, was located there and enrolled students from Oklahoma and the rest of the United States from 1884 to 1980.

CHILTIPIN Creek (TX, Jim Wells Co.) \chil ti PEEN\. From Mexican Spanish *chiltepín,* a variant of *chiltipiquín* 'a type of chili pepper', which in turn is derived from Aztec *chiltecpin,* from *chilli* 'chili pepper' and *tecpin(tli)* 'flea'.

CHIMAL, Valle (NM, San Miguel Co.) \chee MAHL\. From New Mexican Spanish *chimal* 'a homemade shield made of rawhide', from Aztec *chimalli* 'shield'.

CHIMAYÓ (NM, Rio Arriba Co.) \chim uh YOH\. A Spanish approximation of the name of a former Tewa pueblo, said to mean 'good flaking stone'. The word *Chimayosos* in **Chimayosos Peak** (NM, Rio Arriba Co.) \chim uh YOH sohs\ means 'people from Chimayó' in Spanish.

CHINATI Mountains (TX, Presidio Co.) \chi NAH tee\. Perhaps from Mexican Spanish *chanate, zanate* 'blackbird', from Aztec *tzanatl* 'grackle'.

CHINCHONTE, Arroyo (NM, Torrance Co.) \chin CHON tay\. New Mexican Spanish *chinchonte* 'mockingbird,' from Mexican Spanish *sinsonte, zenzontle*, abbreviated from Aztec *centzon-tlahtoleh*, lit. 'having four hundred languages', containing *centzontli* 'four hundred' and *tlahtolli* 'word, language'.

CHINDE (NM, San Juan Co.) \chin DAY, CHEEN dee\. From Navajo *ch'įįdii* 'evil spirit, devil, ghost', also used as an expletive: 'Damn! Hell!'

CHINLE Creek (AZ, Apache Co.; UT, San Juan Co.) \chin LEE\. From Navajo *ch'ínílį* 'it flows outward', from *ch'i* 'outward' and *nílį* 'it flows', because it is at the mouth of Canyon de Chelly.

CHINQUAPIN \CHING kuh pin, SHING kuh pin\. The name of a kind of wild chestnut, recorded in Virginia Algonquian as *chechinquamins* during colonial times. The term occurs as a placename in **Chinquapin** (TX, San Augustine Co.) and **Chinquapin Mountain** (OK, McCurtain Co.).

CHIRICAHUA (AZ, Cochise Co.) \chir i KAH wuh\. The name of an Apache group, from Spanish *Chiricagui*, a name originally applied to mountains in Sonora, Mexico, derived from the Ópata language of that region. The name first appears in Spanish records dated 1588. The Chiricahua Apache people originally lived in southeastern Arizona, southwestern New Mexico, and adjacent areas of Mexico. Under leaders such as Geronimo and Cochise, they were among the last North American tribes to submit to white authorities. Parts of the tribe were exiled to a prison at Fort Marion, Florida, then later moved to Alabama and Oklahoma before most of them were returned to New Mexico. The **Chiricahua Mountains, Chiricahua National Monument,** and the site of the former **Chiricahua Indian Reservation** (1832–76) are all in Cochise County, Arizona.

CHISE (NM, Sierra Co.) \cheez\. Perhaps from Navajo *chíshí*, referring to the Chiricahua Apaches, or perhaps from Chiricahua Apache *chizh* 'firewood'. A possibly related placename is **COCHISE County** (AZ).

CHIULIKAM (AZ, Maricopa Co.) From Tohono O'odham (Papago) *che'ulkam* 'willows', from Tohono O'odham *ce'ul* 'Gooding willow tree, *Salix gooddingii*'.

CHIULI SHA'IK (AZ, Pima Co.) On the Tohono O'odham (Papago) Reservation; from Tohono O'odham *ce'ul ṣa'ik* 'willow brush', with *ṣa'i* 'grass, brush'.

CHOCOLATE Bay (TX, Calhoun Co.) From Spanish *chocolate*, borrowed from Aztec *chocolatl*, a drink made from cacao seeds.

CHOCOLATE Bayou (TX, Brazoria Co.) From Spanish *chocolate,* borrowed from Aztec *chocolatl,* a drink made from cacao seeds. See also **BAYOU**.

CHOCTAW \CHOK tah, CHOK taw\. A people of the southeastern United States, now living in Mississippi and Oklahoma; their language is of the Muskogean family, and the Native name is *chahta,* an unalyzable word. The historical capital of the **Choctaw Nation** is at Tuskahoma, Oklahoma (Pushmataha Co.); the **Choctaw Nation Historical Museum** there was the capitol building of the Choctaw Nation from 1884 to 1907. There is a town called **Choctaw** in Oklahoma (Oklahoma Co.) and a stream called **Choctaw Creek** in Texas (Grayson Co.).

CHOSKA (OK, Wagoner Co.) \CHOOS kuh, CHOH skuh\. In the historic Creek Nation. This Muskogee name, meaning 'post oak' (a species of oak tree), is said to commemorate Polly Postoak, the townsite owner; the Native name is *coskv* 'post oak'.

CHOULIC (AZ, Pima Co.) \CHOO lik\. On the Tohono O'odham (Papago) Reservation; the Tohono O'odham name is *cu:lk,* from *cu:l* 'hipbone'.

CHUCKAHO Creek (OK, Lincoln Co.) Said to be named for a local Sac Indian called *Che-ka-ko.*

CHUCULATE Hill (OK, Wagoner Co.) Taken from the surname of a Cherokee family. There is also a **Chuculate Mountain** in Sequoyah County, Oklahoma.

CHUI TONK (AZ, Pima Co.) \choo ee TAWNK\. From Tohono O'odham (Papago) *cu'i ṭo:nk* 'pollen hill', containing *cu'i* 'pollen'.

CHUI VAYA (AZ, Pima Co.) \choo ee VAH yuh\. From Tohono O'odham (Papago) *cu'i waya* 'pollen well', containing *waya* 'well'.

CHUICHU (AZ, Maricopa Co.) \CHOO choo\. On the Tohono O'odham (Papago) Reservation; from Tohono O'odham *ce:co* 'caves', from *ceho* 'cave'.

CHUKUT KUK (AZ, Pima Co.) \choo koot KUK\. On the Tohono O'odham (Papago) Reservation; from Tohono O'odham *chukuḍ kuk* 'owl hooted', containing *cukuḍ* 'owl'.

CHULA (OK, McCurtain Co.) \CHOO luh\. In the historic Choctaw Nation; the name is perhaps from Choctaw *cholah* 'fox', possibly confused with Spanish *chula* 'pretty'.

CHUSKA Mountains (NM, San Juan Co.) \CHOO skuh\. From Navajo *ch'óshgai* 'white spruce', from *ch'ó* 'spruce tree' and *(sh)gai* '(it is) white'. The **Rio Chusca,** using a Spanish spelling, is in the same county.

CHUTUM VAYA (AZ, Pima Co.) On the Tohono O'odham (Papago) Indian Reservation; contains *waya* 'well'.

CHUWUT MURK (AZ, Pima Co.) On the Tohono O'odham(Papago) Indian Reservation; contains *me:k* 'far'.

CIBECUE (AZ, Navajo Co.) \SIB i kyoo, sib i KYOO\. From Western Apache *deshchíí' bikǫ* 'canyon of horizontally-red-things', combining *deshchíí'* 'horizontally red things' and *kǫ* 'canyon, valley'.

CIBOLA County (NM) \SEE buh luh\. From New Mexican Spanish *cíbola* '(female) buffalo', from earlier *vaca de Cíbola* 'cow of Cíbola', referring to the legendary cities of Cíbola, which were sought by the conquistadores. It has been suggested that *Cíbola* may be derived from the Zuni name for Zuni Pueblo, which is *šiwin'a*. In Arizona, **Cibola National Wildlife Refuge** (La Paz Co.) borders the Colorado River. The related form *cíbolo* 'male buffalo' occurs in the names **Cibolo Mountains** (NM, DeBaca Co.) \SEE buh loh\ and **Cibolo Creek** (TX, Guadalupe Co.) \SEE boh loh\.

CLAREMORE Creek (OK, Rogers Co.) Derived from the name of an Osage leader known in English as *Clermont* or *Claremont*. This may have been derived in turn from an Osage name, *Gra Mo'n,* of unclear meaning. **Lake Claremore** is nearby. In Kansas the same name is written *Claymore* (Montgomery Co.).

CLO-CHEW-TAH Ranch (NM, McKinley Co.) \kloh CHOO tah\. From Navajo *tł'ohchintah* 'among the wild onions', from *tł'ohchin* 'wild onion' (*tł'oh* 'grass', *-chin* 'smell'), and *-tah* 'among'.

CLY Butte (AZ, Navajo Co.) \KLIE\. Perhaps from a Navajo personal name, *tł'aa'í,* originally meaning 'left-handed', now used in the spelling *Cly* as a family name.

COAHOMA (TX, Howard Co.) \kuh HOH muh\. Said to mean 'signal' in an unidentified Indian language.

COBABI (AZ, Pima Co.) \koh BAH bee\. On the Tohono O'odham (Papago) Reservation; from Tohono O'odham *ka:w wahia* 'badger well', containing *ka:w* 'badger' and *wahia* 'well'. Also spelled **Ko Vaya.**

COCHISE County (AZ) \koh CHEES\. Named for a leader of the Chiricahua Apache people. The term may be derived from Chiricahua Apache *chizh* 'firewood', *ko-chizh* 'his firewood', or *ch'izhí* 'the rough one' (from *dich'izh* 'rough'). A possibly related placename is **CHISE** (NM, Sierra Co.).

COCHITI (NM, Sandoval Co.) \KOH chi tee\. A pueblo of the Keresan language family; the Native name is *k'údyîiti,* perhaps meaning 'stone kiva'.

COCOMUNGA Canyon (AZ, Gila Co.) \koh kuh MUNG guh\. Perhaps a transfer name from *Cucamonga* (CA, San Bernardino Co.), derived from Gabrielino *kúkamonga.*

COCONINO County (AZ) \koh kuh NEE noh\. In the forms *Coninas, Cogninas,* and *Cosninos,* this name was applied to the Havasupai people in early Spanish documents; it is from Hopi *kòonina* 'Havasupai person'. The county contains **Coconino National Forest**, surrounding the county seat of Flagstaff. A related placename is **Cosnino** (AZ, Coconino Co.).

COCOPA \KOH koh pah, KOH kuh pah, koh koh PAH, koh kuh PAH\. The name of a Yuman Indian group that originally lived in southeastern California, southwestern Arizona, and adjacent areas of Mexico; its Native name is kokwapá. The English spelling generally used by anthropologists and linguists is *Cocopa;* a varient is *Cocopah.* The **Cocopa Indian Reservation** is in Arizona (Yuma Co.), and **Cocopa Point** (AZ, Coconino Co.) is named for this group.

COLD Spring (AZ, Apache Co.) Corresponds to Navajo *tó sik'az háá̜ lį́* 'cold water flows up and out', containing *tó* 'water', *sik'az* 'it is cold', and *háá̜ lį́* 'it flows up and out'.

COLLE Canyon (NM, Sandoval Co.) \KOH yay\. A spelling of New Mexican Spanish *coye* 'roof-door', from Tewa *k'oyi.* Also written *Coye.*

COMA-A Spring (AZ, Navajo Co.) Said to be from a Navajo pronunciation of Southern Paiute *kamï* 'jackrabbit'.

COMAL (TX, Comal Co.) \KOH mal\. From Mexican Spanish *comal,* referring to the griddle on which tortillas are cooked. The word is from Aztec *comalli.* The **Comal River** runs through the county seat, New Braunfels, where it flows into the Guadalupe River.

COMANCHE \kuh MAN chee\. The name of a Native people of the southern Great Plains states, now living in Oklahoma; their language belongs to the Numic branch of the Uto-Aztecan family. The name is from Ute *kïmánci, kïmáci* 'enemy, foreigner'. In Oklahoma, the name has been given to the town of **Comanche,** (Stephens Co.) and to nearby **Comanche Lake.** The **Comanche Nation** has its headquarters in Lawton, Oklahoma (Comanche Co.), which also is home to the **Comanche National Museum and Cultural Center.** In central Texas, the town of **Comanche** is the seat of **Comanche County,** and there is a **Comanche Peak** in Hood County. There is a **Comanche Creek** in Arizona (Coconino Co.).

COMANCHEROS (NM, San Miguel Co.) \koh mun CHARE ohs\. This Spanish term refers to Hispanic and Anglo entrepreneurs who carried on trade with the Comanches in the nineteenth century.

COMOBABI (AZ, Pima Co.) \koh muh BAH bee\. From Tohono O'odham (Papago) *ko:m wahia* 'hackberry well', containing *ko:m* 'hackberry' and *wahia* 'well'.

CONCHARTY Creek (OK, Wagoner Co.) \kun CHAHR tee\. In the historic Creek Nation; from the name of a Muskogee tribal town, *kvncate,* perhaps from *ēkvnv* 'land, earth' and *cate* 'blood'. **Concharty Mountain** is nearby.

CONCHAS (NM, San Miguel Co.) \KOHN chus\. The Spanish word *conchas* means 'shells', and it is possible that the placename refers to shells used by Indians to make ornaments. However, there may also be confusion with *Conchos,* a term the Spaniards used as a name for an Indian group they encountered in southern New Mexico in the sevententh century.

CONJADA Mountains (OK, Wagoner Co.) \kun JAH duh\. A variant of **CONCHARTY.**

COON. An abbreviation of *raccoon,* the name of a nocturnal mammal (*Procyon lotor*) found throughout the United States. The term was originally recorded by Captain John Smith in Virginia as *aroughcun,* from Virginia Algonquian. As a placename, it is widespread; in some cases it may reflect a secondary meaning in which *coon* is a derogatory term for an African American. There are several streams called **Coon Creek** in Oklahoma (for example, in Osage Co.) and Texas (for example, in Briscoe Co.).

COOSA (OK, Latimer Co.) \KOO suh\. In the historic Choctaw Nation; the probable origin is Choctaw *kǫshak* 'cane' (the plant).

COOWEESCOOWEE (OK, Rogers Co.) This town was the seat of Cooweescoowee District of the Cherokee Nation during the period 1856–1907. The word is said to have been the personal name of a prominent Cherokee man known in English as John Ross. The name may come from Cherokee *gu'wi'sgu'wi',* an onomatopoeic term for a large bird, possibly a snipe with yellow legs.

COO-Y-YAH (OK, Mayes Co.) \KYOO ee yah\. In the historic Cherokee Nation; said to be from Cherokee *kuwahiyi* 'mulberry grove', containing *kuwa* 'mulberry'.

COPAN (OK, Washington Co.) \KOH pan\. From Spanish *Copán,* the name of a town in Honduras famous for its Mayan ruins. Adjacent to the Oklahoma town are **Copan Lake** and **Copan Wildlife Management Area.**

CORRUMPA (NM, Union Co.; OK, Cimarron Co.) \kuh RUM puh\. Said to represent an Indian word meaning 'wild, isolated', but the language of origin is unidentified. Its similarity to Spanish *corrumpir* 'to corrupt', however, is striking. The stream flows eastward and becomes the headwaters of the Beaver River (North Canadian River).

COSNINO (AZ, Coconino Co.) \kohz NEE noh\. From Hopi *kòonina* 'Havasupai person'. A related placename is **COCONINO County** (AZ).

COTTONWOOD Tank (AZ, Coconino Co.) Corresponds to Navajo *t'iis sitání* 'the place where the cottonwood is lying', containing *t'iis* 'cottonwood tree(s)' and *sitą́* 'it is lying'.

COUSHATTA Creek (TX, Colorado Co.) \koo SHAT uh\. Coushatta is the name of a Muskogean people, formerly living in Alabama; descendants of the Coushatta now live on the **Alabama-Coushatta Reservation** near Livingston, Texas (Polk Co.) and in the **Alabama-Quassarte Tribal Town** at Henryetta, Oklahoma (Okmulgee Co.). The name is from a Choctaw word meaning 'white cane', from *kǫshak* 'cane' and *hata* 'white'. A related form of the name is **Koasati** (AL, Elmore Co.) \koh uh SAH tee\; this form is the one now most commonly used by anthropologists and linguists.

COW Springs (AZ, Coconino Co.) Corresponds to Navajo *béégashii bito'* 'cow's spring', containing *béégashii* 'cow', a borrowing from the Spanish plural *vacas* 'cows'.

COWETA (OK, Wagoner Co.) \koh WEE tuh, kou EE tuh\. In the historic Creek Nation; a transfer name from Coweta County in Georgia. The term is derived from Muskogee *kvwetv,* the name of a tribal town. Between 1867 and 1907, **Coweta District** was part of the Creek Nation in Oklahoma.

COWLIC (AZ, Pima Co.) \KOU lik\. From Tohono O'odham (Papago) *kawlĭk* 'hilly'.

COYE Canyon (NM, Sandoval Co.) \KOH yay\. New Mexican Spanish *coye* 'roof-door', from Tewa *k'oyi.* Also written **Colle.**

COYOTE \kah YOH tee, KIE oht\. An animal of the dog family, found over much of North America. The word is from Spanish *coyote,* from Aztec *coyotl.* As a placename, it is widespread, especially in the western United States—for example, in the names **Coyote Hill** (OK, Canadian Co.) and **Coyote Corner** (TX, Andrews Co.). In New Mexico, **Coyote Canyon** (McKinley Co.) corresponds to Navajo (Athabaskan) *mą'ii téeh yítłizhí* 'where the coyote fell into deep water', containing *mą'ii* 'coyote', *téeh* 'deep water', and *yítłizh* 'he fell in'.

CREEK \kreek\. The name of a major Indian group of the southeastern United States belonging to the Muskogean language family. The Creeks formerly lived mostly in Georgia and Alabama; now they are in the Creek and Seminole Nations of Oklahoma and the Seminole community of Florida. The name *Creek* was abbreviated from *Ochese Creek Indians,* referring to a stream in Georgia where the group was first contacted by English traders; that stream is now called the Ocmulgee River. The Creek people are also called **Muskogee** or **Muscogee,** a word of unclear derivation, and the name *Muskogee* is used elsewhere in this book in reference to the people and the language. The **Muscogee (Creek) Nation of Oklahoma** has its historic capital at Okmulgee (Okmulgee Co.). The **Creek Council House Museum** in Okmulgee served as the capitol building of the Creek Nation until 1906.

CREKOLA (OK, Muskogee Co.) \kree KOH luh\. In the historic Creek Nation; coined from the word *Creek* plus the Anglicizing suffix *–ola,* as in *Crayola* and *granola.*

CROSS CANYON Trail (NM, Apache Co.) From Navajo *ałnaashii ha'atiin* 'a trail that goes up on opposite sides', containing *ałnaashii* 'on opposite sides', *ha-* 'up', and *atiin* 'trail, road'.

CROW Trail (AZ, Apache Co.) From Navajo *gáagii haayáhí* '(where) the crow ascended', containing *gáagii* 'crow'.

CRUMBLED HOUSE (NM, San Juan Co.) Corresponds to Navajo *kin náázhoozhí* 'sliding house', containing *kin* 'house'.

CRYSTAL (NM, San Juan Co.) Corresponds to Navajo *tóniłts'ílí* 'sparkling water', from *tó* 'water' and *niłts'ílí* 'sparkling'.

CUATE Canyon (NM, Socorro Co.) \KWAH tay\. This Mexican Spanish word means 'twin', from Aztec *coatl,* which also means 'snake'. The plural **Cuates** also occurs as a placename (NM, Union Co.).

CUBA (NM, Sandoval Co.) The name of the island nation in the Caribbean Sea became a popular placename in the United States at the time of the Spanish-American War, at the end of the nineteenth century. The term is said to be from the Taino name *Cubanacan,* referring to the people who occupied the island. Spanish *cuba,* however, can also mean 'trough, tank', and this may be the source of the New Mexico placename, which dates to before the Spanish-American War. There is also a **Cuba** in Texas (Johnson Co.).

CULLEOKA (TX, Collin Co.) Perhaps a transfer name from Tennessee; from Choctaw *kali-oka,* lit. 'spring water'.

CUMA, Arroyo (NM, Santa Fe Co.) \KOO muh\. Perhaps a Spanish spelling of Tewa *t'oma,* a name for the promontory called Red Hill.

CUMARO Canyon (AZ, Pima Co.) \koo MARE oh\. From local Spanish *cumero* 'hackberry tree', from Tohono O'odham (Papago) *ko:m* 'hackberry tree'. A related form is **Cumero Canyon** (AZ, Santa Cruz Co.).

CUNDIYO (NM, Santa Fe Co.) \koon DEE yoh\. From the Tewa placename *kųdiyo.* A variant spelling is **Cundayo.**

CUNNEO TUBBY Creek (OK, Latimer Co.) \kun uh TUB ee\. In the historic Choctaw Nation; perhaps from the war name of a Choctaw man, of the form *kaniya-t-abi* '(one who) went away and killed', containing *kaniya* 'to go away' and *abi* 'to kill'.

CUSSETA (TX, Cass Co.) Perhaps a transfer name from Alabama; from Muskogee *kvsehtv,* the name of a tribal town. There is also a **Cussetah Creek** in Oklahoma (Okmulgee Co.), flowing into the Deep Fork River.

CUYAMUNGUE (NM, Santa Fe Co.) \koo yuh MOONG gay\. From Tewa *k'uyemuye* 'where they threw down the stones', containing *k'u* 'stone' and *yemu* 'to throw things down'.

D

DACOMA (OK, Woods Co.) \duh KOH muh\. From a combination of *Dakota,* of Siouan origin, and *Oklahoma,* of Choctaw origin.

DADASOA Spring (NM, McKinley Co.) Perhaps from Navajo *dah ast'os* 'cone-shaped'.

DAHLONEGAH Mountain (OK, Adair Co.) \dah LON uh guh, duh LON uh guh\. In the historic Cherokee Nation; the name is a transfer from a gold rush town of the 1830s in Georgia (Lumpkin Co.), in the original Cherokee territory. The name is from Cherokee *adel dalonige* 'gold', lit. 'money-yellow'.

DAKAI Well (NM, McKinley Co.) \DAH kie\. Perhaps from Navajo *ndaakai* 'they wander around', referring to Spanish Americans. A related placename is **NOKAI Canyon** (AZ, Navajo Co.).

DAKOTA Spring (NM, Rio Arriba Co.) \duh KOH tuh\. The word *Dakota* refers to a group of peoples that lived in the northern Great Plains states and formed a major branch of the Siouan language family. The source of the English term, which appears in the state names *North Dakota* and *South Dakota,* is a word often translated as 'allies'; related forms are Dakota *dakhóta* 'friendly' and Lakhota *kholá.*

DEBEBEKID (AZ, Navajo Co.) \di BAY buh kid\. From Navajo *dibé be'ek'id,* 'sheep lake', containing *dibé* 'sheep'.

DE CHELLY, Canyon (AZ, Apache Co.) \duh SHAY, dee SHAY\. The Spanish spelling *Chelly* represents Navajo *tsé-yi'* 'canyon' (lit. 'rock-inside'). In **Canyon de Chelly National Monument**, surrounded by the Navajo Nation, remains from five periods of Native American culture are found, dating from as early as 2500 B.C.—Archaic, Basketmaker, early Pueblo, Hopi, and Navajo. Pictographs on the canyon walls date from the earliest occupation. Visitors can hire authorized Navajo guides to make trips into the canyon.

DELAWARE \DEL uh wair\. This word, originally the surname of the British nobleman Thomas West, Lord De la Warr, was first applied as a placename to the Delaware River and to what later came to be the state of Delaware. The British also applied the name to an Indian people, speakers of an Algonquian language, who occupied much of the Delaware River valley of Pennsylvania in colonial times, as well as all of New Jersey and the area around New York City. Originally a toponym, the term frequently refers to the Indian people rather than to the river or the state. The town of **Delaware** in Oklahoma (Nowata Co.) was the seat of Delaware District in the Cherokee Nation during the period 1840–1907. There is also a **Delaware Creek** in Oklahoma (Atoka Co.). At present, two organized groups of Delaware Indians exist in Oklahoma: the **Delaware Tribe of Indians** in Bartlesville (Washington Co.) and the **Delaware Nation** in Anadarko (Caddo Co.). In Texas, a **Delaware River** flows into the Pecos River in Eddy County.

DELNAZINI Spring (NM, San Juan Co.) \del nah ZEE nee\. From Navajo *déél nazíní* 'cranes standing'.

DEL SHAY Basin (AZ, Gila Co.) \del SHAY\. Named for a Western Apache leader who died in 1874. His name was *de'iłché'é* 'red ant'.

DENNEHOTSO (AZ, Apache Co.) \day ni HOH tsoh, den ay HOH tsoh\. From Navajo *deinihootso* 'yellow area extending upward'.

DESHGISH (AZ, Navajo Co.) \DESH gish\. From Navajo *deeshgizh* 'gapped butte'.

DEZA Bluffs (NM, McKinley Co.) \duh ZAY\. From Navajo *deez'á* 'point, promontory, elongated ridge'.

DIBE CHAA Valley (AZ, Apache Co.) \di BAY chah\. Perhaps from Navajo *dibé ntsaa,* 'big sheep' (lit. 'sheep-big').

DILKON (AZ, Navajo Co.) \DIL kon, dil KON\. This is a shortening of Navajo *tsé-zhin dilkǫǫh,* 'smooth black rock' (lit. 'rock-black is-smooth').

DINNE Mesa (AZ, Apache Co.) \di NAY\. From Navajo *diné* 'human being, Navajo'.

DINNEBITO Wash (AZ, Coconino Co.) \din ay BEE toh\. From *diné bito'* 'Navajo's spring'.

DINNEHOTSO (AZ, Navajo Co.) \di NAY hot soh, di nay HOT soh\. From Navajo *diné hotso* 'upper end of meadow'.

DOCK (AZ, Pinal Co.) \dahk\. From O'odham *dak* 'sitting'.

DO HA HI BITOH (AZ, Navajo Co.) \doh HAY bi toh\. From Navajo, probably *doo hahí bitó'* 'slow man's water', containing *doo hahí* 'slowly'.

DOWOZHIEBITO Canyon (AZ, Navajo Co.) From Navajo *diwózhii bii' tó* 'spring in the greasewood', from *diwózhii* 'greasewood', *bii* 'in it', and *tó* 'water'.

DZIL DASHZHINII (AZ, Navajo Co.) \dzil DAHSH ji nee\. From Navajo *dził dashzhinii* 'black mountain', containing *dził* 'mountain'.

DZIL NDA KAI Mountain (NM, McKinley Co.) \dzil DAY kie\. Perhaps from Navajo *dził ndaakai* 'mountains walk around', containing *ndaakai* 'they walk around'.

DZILNTSAHAH Summit (NM, San Juan Co.) \dzil nuh TSAH hah\. From Navajo *dził ntsaa* 'big mountain', containing *ntsaa* 'big'.

DZILTH-NA-O-DITH-HLE (NM, San Juan Co.) \dzil nah OH dith lee\. From Navajo *dził ná'oodiłii* 'Huerfano Mountain', lit. 'the mountain around which people move', containing *dził* 'mountain' and *ná'oodiłii* 'people circle around'.

ECHO Cave (AZ, Navajo Co.) Corresponds to Navajo *tséyaa hodiits'a'í* 'rock under which there is an echo'.

ECLETO (TX, Karnes Co.) \i KLEE toh\. Said originally to have been *Cleto,* an Indian name; then turned by Spaniards into *El Cleto* and later by Anglos into its present form.

ECONTUCHKA (OK, Pottawatomie Co.) From the Muskogee village name *ēkvntvckv,* meaning 'surveyed line, boundary'.

EDA HUGKAM SWADAG (AZ, Pima Co.) From Tohono O'odham (Papago) *eḍa hugkam swadagĭ* 'half wet', containing *wadagĭ* 'wet'.

EMACHAYA Creek (OK, Haskell Co.) In the historic Choctaw Nation; the creek is a tributary of the Canadian River and is named for one of the earliest Choctaws to settle in this region. Soil scientists have given the name *Emachaya series* to a type of soil typical of this area.

EMIKA (AZ, Pima Co.) \EE mik uh\. From Tohono O'odham (Papago) *i:miga* 'kinship terms'.

ERIE (OK, Kay Co.) \EER ee\. The name is that of an Iroquoian people originally living around Lake Erie in the Great Lakes region; they were largely dispersed during expansion of the white population in the eighteenth and nineteenth centuries. The English name is from French *Erié,* an abbreviation of the Native term *Rhiienhonons,* 'people of the racoon'. The name **Lake Erie** occurs in New Mexico (Cibola Co.).

ESCHITA (OK, Tillman Co.) \ESH i tah\. Named for a nineteenth-century Comanche leader; the Native term is *isa-tai,* lit. 'wolf-vulva'.

ESKIMINZIN Wash (AZ, Pinal Co.) \es kuh MIN zin\. Named for a Western Apache leader called *haškế baanʒin* 'they stand in line for him who is angry', containing *haškế* 'angry'.

ESKIMO \ES ki moh\. This term refers to an ethnic and linguistic grouping of peoples who live in Arctic areas and speak a distinctive set of language varieties. They fall into two linguistic groups. One, the Innuit or Iñupiaq, lives in areas bordering the Arctic Ocean from Greenland westward, across Canada and into Alaska, as far as the Bering Strait. The other group, the Yupik, lives in southwestern Alaska in areas bordering the Chukchi Sea and the Gulf of Alaska. The word *Eskimo* has its first English attestation in 1584, in the form *Esquimawes*; this was later written *Esquimaux* and was used as both a singular and a plural in both French and English. The term *Eskimo* became more common after the mid-nineteenth century. It is likely that the word entered both French and

English from Montagnais, an Algonquian language in which early names for the Eskimos such as *aiachkimeou* and *aiachtchimeou* were recorded. Some Algonquian languages refer to the Eskimos by a term that means 'eaters of raw meat' (for example, Eastern Ojibwa *ashkipook*); and for this reason some Eskimo people prefer to be referred to by other terms. In Canada, official usage has adopted the term *Inuit* for the people and *Inuktitut* for the language. In Alaska, the term *Iñupiaq* refers to the people on the Arctic coast north of the Bering Strait, and the term *Yupik* to those in southeastern Alaska. Eskimo speakers on the Gulf of Alaska are more specifically referred to as *Alutiiq,* a term that derives historically from the fact that Russians identified them with Aleuts, who in fact speak a separate language. As a placename, **Eskimo** is attached to many places in Alaska and to a few elsewhere, such as **Eskimo Tank** near Shiloh Draw in New Mexico (Otero Co.).

ETOI KI (AZ, Pima Co.) \ee toi KEE\. From Pima *i'itoi ki* 'house of I'itoi', referring to the culture hero so named (*ki* means 'house').

ETOWAH Church (OK, Cleveland Co.) \ET oh wah, ET uh wah\. Said to be named for a Cherokee settlement called *Itawa,* in Floyd County, Georgia; however, the word may be a borrowing from Catawban *itá:wą* 'longleaf pine tree'.

EUCHA, New (OK, Delaware Co.) \YOO chuh\. A variant of **EUCHEE**. Near the town of New Eucha are **Lake Eucha** and **Lake Eucha State Park**. See also **YUCHI.**

EUCHEE \YOO chee\. The name of a Native people long assocated politically with the Muskogees but speaking a separate language. They originally lived in Georgia and Alabama, but their descendants now form the **Yuchi (Euchee) Tribe**, a division of the Muscogee Creek Nation living at Sapulpa, Oklahoma (Creek Co.). **Euchee Creek** flows nearby. The Muskogee form of the name is *yocce.*

EUFAULA (OK, McIntosh Co.) \yoo FAH luh, yoo FAW luh\. A town in the traditional Creek Nation; the term is from Muskogee *yofalv,* the name of a traditional tribal town in Alabama. In the period 1867–1907, **Eufaula District** was part of the Creek Nation. In Oklahoma, **Eufala Lake** lies on the Canadian River in McIntosh, Pittsburg, and Haskell counties; the adjacent **Eufala Wildlife Management Area** is in McIntosh County.

F

FACE Rock (AZ, Apache Co.) A translation of Navajo *tsé binii'í* 'rock's face', containing *tsé* 'rock'.

FAR SPIRAL Canyon (AZ, Apache Co.) A translation of Navajo *nízaa'aliwozh nástł'ah* 'far spiral cove'; containing *nástł'ah* 'cove'.

FATHER Rock (AZ, Apache Co.) A translation of Navajo *tsé bazhei* 'rock's father'.

FENCED UP HORSE Valley (NM, McKinley Co.) A translation of Navajo *łį́į́ biná'asht'ih* 'horse enclosed-in-fence', containing *łį́į́* 'horse'.

FIVE CIVILIZED TRIBES. This was the term used to refer to the Cherokee, Chickasaw, Choctaw, Creek, and Seminole nations, most of whose people were uprooted from their homes in the southeastern states after the 1830 Indian Removal Act and resettled in what was then called Indian Territory, now the eastern part of Oklahoma. The **Five Civilized Tribes Museum** in **MUSKOGEE** (Muskogee Co.), Oklahoma, documents their history.

FIXICO Cemetery (OK, Seminole Co.) \FIK si koh\. This name represents a common surname used by Muskogee families. It is from the male war name *fekseko* 'heartless'.

FOREST Lake (AZ, Navajo Co.) A translation of the Navajo *tsinyi' be'ek'id* 'lake inside the trees' (lit. 'trees-inside lake'), containing *be'ek'id* 'lake'.

FOX The name of this Algonquian people, closely linked to the Sac (or Sauk), is a translation of French *Renards,* which in turn may be a translation of an Iroquoian term meaning 'red-fox people'. Members of the unified Sac and Fox tribe now live in several states; the largest group is the **Sac and Fox Nation,** with headquarters in Stroud, Oklahoma.

FU GYO Spring (AZ, Coconino Co.) \FOO gyoh\. From Havasupai *'fú gyo* 'willow place'.

FUZZY Mountain (NM, McKinley Co.) A translation of Navajo *dził ditłʼoii* 'fuzzy mountain', containing *dził* 'mountain'.

G

GAMBLER'S Spring (NM, San Juan Co.) A translation of Navajo *nááhwiłbįįhí bitooh,* referring to a legendary figure who is called the Gambler.

GAP Rock (AZ, Apache Co.) A translation of Navajo *tsé' nídeshgizh* 'rock gap', containing *tsé'* 'rock'.

GATAGAMA (AZ, Coconino Co.) From the surname of a Havasupai family.

GERONIMO (AZ, Graham Co.) \juh RON i moh\. From the name of the famous nineteenth-century Chiricahua Apache leader. The term is from Spanish *Gerónimo, Jerónimo,* corresponding to English *Jerome.* As a placename, **Geronimo** also occurs in New Mexico (San Miguel Co.), Oklahoma (Comanche Co.), and Texas (Guadalupe Co.).

GILA \HEE luh\. The **Gila River** starts as a stream in New Mexico and flows west through Arizona, through the town of **Gila Bend** (Maricopa Co.), to join the Colorado River near Yuma. The name occurs in early Spanish records as *Xila, Gila, Hyla,* and so forth. It was probably borrowed from an Indian language, but the source is now unknown. The first farms around Gila Bend were established by Father Eusebio Kino in 1699. New Mexico has a town called **Gila** (Grant Co.), as well as the **Gila National Forest** (Grant, Catron, and Sierra Cos.).

GIUSEWA (NM, Sandoval Co.) \JOO suh wah\. From Jemez (Towa) *giųsewa* 'hot place', referring to hot springs.

GOING SNAKE (OK, Adair Co.) Named for a Cherokee leader. The town was the seat of **Going Snake District** in the Cherokee Nation during the period 1840–1907.

GOPHER Canyon (AZ, Apache Co.) A translation of Navajo *na'azísí nástłah* 'gopher cove', containing *na'azísí* 'gopher'.

GOSSIP Hills (NM, McKinley Co.) A translation of Navajo *aseezí* 'gossip, rumor, news'.

GOTEBO (OK, Kiowa Co.) \GOH dee boh\. Named for a leader of the Kiowa people, *Qó:dèbòbòn* 'mink hat'.

GRAN QUIVIRA (NM, Torrance Co.) \grahn kee VERE uh\. A Pueblo Indian ruin in Salinas Pueblo Missions National Monument. The Spanish name, meaning 'great Quivira', refers to a legendary city sought by the Spanish explorer Coronado, now thought to have been in Kansas. The Native American origin of the name is now unknown.

GRAY HORSE (OK, Osage Co.) Named for an Osage man whose Native name was *kkáwa xó:ce,* lit. 'horse gray'. The Osage word *kkáwa* 'horse' is a borrowing from Spanish *caballo.*

GRAY WHISKERS Butte (AZ, Navajo Co.) A translation of Navajo *dághaa' łibáí* 'gray whiskers', containing *dághaa'* 'whiskers'.

GREASEWOOD (AZ, Navajo Co.) This name for a native shrub is an abbreviated translation of Navajo *diwózhii bii' tó* 'a spring in the greasewood', containing *diwózhii* 'greasewood'.

GREEN COTTONWOOD Canyon (AZ, Apache Co.) An abbreviated translation of Navajo *t'iis ndiitsoi ntista'ah* 'yellow-green cottonwoods extending downward cove', containing *t'iis* 'cottonwood'.

GU ACHI (AZ, Pima Co.) \goo AH chee\. On the Tohono O'odham (Papago) Reservation; from Tohono O'odham *ge aci* 'big narrow'. Also written **Gu-Achi.**

GU CHUAPO (AZ, Pima Co.) \goo CHOO poh\. On the Tohono O'odham (Papago) Reservation; from Tohono O'odham *ge ce:po,* lit. 'big mashing-pit'.

GU KOMELIK (AZ, Pima Co.) \goo KOH muh lik\. From Tohono O'odham (Papago) *ge komlik,* lit. 'big flats'.

GU KUI CHUCHG (AZ, Pima Co.) \goo koo ee CHOOCHK\. From Tohono O'odham (Papago) *ge kui cu:ck,* lit. 'big mesquite stand', from *kui* 'mesquite' and *cu:ck* '(plural) to stand'.

GU OIDAK (AZ, Pima Co.) \goo OI dahk\. From Tohono O'odham (Papago) *ge oidag* 'big field', containing *oidag* 'field'.

GU VO (AZ, Pima Co.) \goo VOH\. From Tohono O'odham (Papago) *ge wo'o,* lit. 'big pond', containing *wo'o* 'pond'.

GUACAMALLA Canyon (NM, Sandoval Co.) \gwah kuh MAH yuh\. A variant spelling of Spanish *guacamaya* 'macaw, a type of parrot', originally borrowed from a language of the West Indies. The Spanish masculine form occurs in **Guacamayo Historic Site** \gwah kuh MAH yoh\, also in Sandoval County.

GUACHE, El (NM, Rio Arriba Co.) \WAH chay\. Also written **El Guacho;** perhaps an abbreviation of **GUACHUPANGUE.**

GUACHUPANGUE (NM, Rio Arriba Co.) \gwah choo PAHNG gay\. Earlier written *Guachepanque;* a Spanish adaptation of Tewa *p'otsip'ą'nge* 'down at the mud-string place'.

GUAJALOTE, Arroyo (NM, San Miguel Co.) \gwah huh LOH tee\. From Mexican Spanish *guajolote* 'turkey', derived in turn from Aztec *huehxolotl* 'male turkey'.

GUAJE Canyon (NM, Santa Fe Co.) \GWAH hay\. From Spanish *guaje* 'gourd', derived in turn from Aztec *huaxin.*

GUAYULE Creek (TX, Brewster Co.) \gwah YOO lee\. The name is that of a desert plant that is a source of rubber; it is a Mexican Spanish word from Aztec *cuauh-olli* 'wood-rubber'.

GUEVAVI Canyon (AZ, Santa Cruz Co.) \gay VAH vee\. This name was given to the first Spanish mission founded in Arizona, in 1701. Its Native American source has not been established.

GUIQUE, EL (NM, Rio Arriba Co.) \GEE kay\. From a Spanish spelling, *Quigiii,* from Keresan *dyîiwi,* the name for Kewa (Santo Domingo) Pueblo in some Keresan dialects (other than that of Kewa Pueblo itself, in which the name is *dyîiwa*). Probably from a Tewa placename, otherwise unidentified. A probably related placename is **QUIGUI** (NM, Sandoval Co.).

GUKAWO KI (AZ, Pima Co.) From Tohono O'odham (Papago) *gegewho ki:,* lit. 'wildcats' house', from *gewho* 'wildcat' and *ki:* 'house'.

GURLI PUT VO (AZ, Pima Co.) \GUR li put VOH\. On the Tohono O'odham (Papago) Reservation, from Tohono O'odham *kelibaḍ wo'o* 'dead man's pond', containing *wo'o* 'pond'.

H

HA HO NO GEH Canyon (AZ, Coconino Co.) This Navajo name, recorded around 1858, is said to mean 'too many washes', but the derivation is unclear. The current Navajo name for the place is said to be *nahonsheeshjééł* 'rough rocky canyon'.

HAIVAN VAYA (AZ, Pima Co.) \hie vahn VAH yah\. Tohono O'odham (Papago) *haiwañ wahia* 'cow well', containing *haiwañ* 'cow, cattle' and *wahia* 'well'.

HAIVANA NAKYA (AZ, Pima Co.) \hie vah na NAHK yah\. Tohono O'odham (Papago) *hawañ naggia* 'crow hang', containing *hawañ* 'crow' and *naggia* 'hanging'.

HAKATAI Canyon (AZ, Coconino Co.) \HAH kuh tie\. Perhaps the Havasupai name for the Colorado River, meaning 'big water', containing *ha* 'water' and *tay* 'big'.

HALI MURK (AZ, Pima Co.) \hah lee MURK\. Tohono O'odham (Papago) *ha:lĭ mek* 'burnt squash', from *ha:lĭ* 'squash' and *mek* 'burnt'.

HAMIDRIK (AZ, Coconino Co.) This is the surname of a Havasupai family and is said to mean 'nighthawk'. Also written **Hamidreek.**

HANEY (OK, Seminole Co.) \HAY nee\. Said to have been named for a Seminole Indian, the Reverend Willie Haney.

HANO (AZ, Navajo Co.) \HAH noh\. This community on First Mesa in the Hopi Reservation, also called **Tewa Village,** is occupied by descendants of Tewa Indians from the Rio Grande Valley of New Mexico who took refuge with the Hopis during the days of Spanish colonial rule. The people of Hano, also called "Hopi-Tewas" and "Arizona Tewas," speak both Hopi (Uto-Aztecan) and Tewa (Kiowa-Tanoan). The name *Hano* is from Hopi *haano,* borrowed from Tewa *tháanu,* referring to an abandoned pueblo in New Mexico; this is the source of the word *Tanoan* in the name for the language family to which Tewa belongs.

HARCUVAR (AZ, La Paz Co.) \HAHR kuh vahr, hahr kuh VAHR\. From Mojave *ha-kavaar,* shortened from *'ahá kaváar* 'there's no water', containing *'ahá* 'water' and *kaváar* 'be not'.

HARD GROUND Canyon (NM, McKinley Co.) Corresponds to Navajo *ni' hótłizí* 'hard ground', from *ni'* 'earth' and *hótłiz* '(area) to be hard'.

HARJO (OK, Pottawatomie Co.) \HAR joh\. A Muskogean surname, originally a title, *haco,* sometimes translated as 'crazy'.

HARQUA (AZ, Maricopa Co.) \HAR kwah\. This seems to contain Mojave *'ahá, ha-* 'water', but otherwise its derivation is unclear. Possibly it is an abbreviated form of the Mojave name that appears in English as **Harquahala Mountain** \HAR kwuh hah luh, hahr kwuh HAH luh\, which is probably from Mojave *'avií hakwahél,* the name of a mountain range. The name seems to mean 'rock where water flows', from *'avií* 'rock', *ha-* 'water', and the verb root *–hel,* which apparently means 'to flow'.

HASBIDITO Creek (AZ, Apache Co.) \hahs bi DEE toh\. From Navajo *hasbídí tó,* 'turtle-dove spring', containing *hasbídí* 'turtle-dove'.

HASHAN CHUCHG (AZ, Pima Co.) \hah sahn CHOOCHK\. Tohono O'odham (Papago), meaning 'standing saguaro cactus', from *ha:ṣañ* 'saguaro' and *cu:ck* 'standing'.

HASSAYAMPA (AZ, Maricopa Co.) \has uh YAM puh, hah suh YAHM puh\. Perhaps from Yavapai *'ha syáamvoo,* containing *'ha* 'water'; it may mean something like 'smooth running water'. The alternate form **Hassiamp** \HAS uh yamp\ also occurs.

HATOBI (OK, McCurtain Co.) \HAH toh bee\. In the historic Choctaw Nation; perhaps from a Choctaw war name, *hattak-abi* 'person killer', from *hattak* 'person' and *abi* 'to kill'.

HAUANI Creek (OK, Marshall Co.) \huh WAH nee\. In the historic Chickasaw County; from Chickasaw *hayowani'* 'worm'. Also written **Hauwani.**

HAULAPAI Island (AZ, Mohave Co.) \WAH luh pie\. From the same source as **HUALAPAI.**

HAUWANI Creek (OK, Marshall Co.) \huh WAH nee\. A variant of **HAUANI.**

HAVANA (TX, Hidalgo Co.) \huh VAN uh\. From Spanish *La Habana,* the capital of Cuba; originally from a Native American language of the West Indies, now extinct.

HAVASU Canyon (AZ, Coconino Co.) \HAV uh soo\. From Havasupai *havasu* 'blue water', containing *vasu* 'blue'. The name of the **Havasupai Indian Reservation** \hav uh SOO pie\ and of the tribe means 'Havasu people'. The area is inaccessible to automobiles; visitors must enter on foot, on horseback, or by helicopter. The name of **Lake Havasu** on the Colorado River (AZ, Mohave Co.) is from Mojave *havasúu* 'blue'; the lake was formed by the impoundment of Parker Dam in 1938. In 1968, London Bridge was dismantled in England and reconstructed at Lake Havasu City, stone by stone.

HA WHI YALIN Wash (AZ, Navajo Co.) The Navajo name is said to be *hahoyílíní,* translated as 'up well flow'.

HAWIKUH (NM, Cibola Co.) \HAH wi koo\. This archeological site was once a Zuni village; Spanish explorers first encountered it in 1539. The name was later written **AGUICO,** and the Zuni tribe currently spells it **Hawikku.**

HEAD ROCK (AZ, Apache Co.) Corresponds to Navajo *tsé bitsii'í* 'rock's head', containing *tsé* 'rock' and *bitsii'* 'its head'.

HELISWA (OK, Seminole Co.) \huh LIS wuh\. From Muskogee *heleswv* 'medicine'.

HELOTES (TX, Bexar Co.) \hel OH tis\. From Mexican Spanish *elotes* 'ears of sweet corn', the plural of *elote* 'sweet corn', from Aztec *elotl.*

HIAWATHA \hie uh WAH thuh\. This name became popular as a placename because of Longfellow's popular narrative poem *Hiawatha.* Although Longfellow took most of the features of his story from the Ojibwas, an Algonquian people of the Great Lakes area, he borrowed the name of his protagonist from a legendary hero of the Iroquois, called *hayę́hwàtha.* There is a town called **Hiawatha** in Oklahoma (Le Flore Co.) and a **Lake Hiawatha** in Texas (Bandera Co.).

HICKIWAN (AZ, Pima Co.) \HIK i wahn\. From Tohono O'odham (Papago) *hikiwañ* 'jagged cut'.

HICKORY. The name of the tree and of its edible nuts, native to the eastern United States, is shortened from Virginia Algonquian *pocohicora,* a milky drink made from hickory nuts. There are streams called **Hickory Creek** in Texas (Houston Co.) and Oklahoma (Love Co.).

HICO (TX, Hamilton Co.) \HIE koh\. The origin of this name is unclear, but it may be related to the name of a Caddoan Indian group, written as *Hico, Hueco,* or, now, as in the placename **WACO** (TX, McLennan Co.).

HILLABEE (OK, McIntosh Co.) \HIL uh bee\. From *helvpe,* the name of a Muskogee tribal town in Alabama.

HIMMONAH (OK, Garvin Co.) \HIM uh nah\. Perhaps abbreviated from an expression meaning 'new place' (Chickasaw *-himona'* means 'new').

HITCHITA (OK, McIntosh Co.) \HICH i tuh, hi CHEE tuh\. This name refers to a Muskogean people and language of Georgia and to their tribal town in the Creek Confederacy. The language is still spoken among the Seminole community of Florida. Derivation of the term is from Muskogee *hecete* or *hecetv,* the name of the tribal town; linguists use the spelling *Hitchiti.*

HIWASSEE, Lake (OK, Oklahoma Co.) A transfer of the name of the Hiwassee River in Georgia, North Carolina, and Tennessee; from Cherokee *ayuhwasi* 'meadow'.

HO HO KAM Village (AZ, Maricopa Co.) \hoh hoh KAHM, hoh HOH kum, HOH hoh kum\. Archaeologists apply the name *Hohokam* to the ancient cultures of southern Arizona that are known for their large systems of irrigation canals. The name is derived from Tohono O'odham (Papago) *huhugkam* 'those who have perished', from *huhug* 'to perish'. Ho Ho Kam Village is the archaeological site of an ancient Hohokam village. There is a **Ho Ho Kam Mobile Village** today in Coolidge, Arizona (Pima Co.).

HOA MURK (AZ, Pima Co.) \hoh uh MURK\. From Tohono O'odham (Papago) *hoa mek* 'burnt basket', combining *hoa* 'basket' and *mek* 'burnt'.

HOCHATOWN (OK, McCurtain Co.) \HOH chuh toun\. The first element may be from Choctaw *hachcha* 'river'. **Hochatown State Park** is in the same county.

HOCHUBBEE (OK, Le Flore Co.) \hoh CHUB ee, HOCH uh bee\. From Choctaw *hochokbi* 'cave, cavern'.

HOGAN \HOH gahn, HOH gun\. A term for a traditional Navajo house; the Native word is *hooghan* 'someone's home'. This is probably the origin of placenames such as **Hogan Ridge** (AZ, Coconino Co.), **Hogan Well** (NM, McKinley Co.), and **Hogan Mesa** (NM, San Juan Co.). Elsewhere, however, *Hogan* is more likely to represent an English surname of Irish origin. See also **HOGANSAANI Spring.**

HOGANSAANI Spring (AZ, Apache Co.) \hoh gun SAH nee\. The name is supposedly Navajo for 'lone hogan', but more correct Navajo would be *sahdii hooghani*. The English word *hogan,* referring to a traditional Navajo house, is from Navajo *hooghan* 'someone's home'.

HOI OIDAK (AZ, Pima Co.) \hoi OI dahk\. From Tohono O'odham (Papago) *ho'i oidak* 'thorn field', containing *ho'i* 'thorn' and *oidak* 'field'.

HOMINY. Refers to hulled corn from which the bran and germ have been removed by soaking the corn in lye or by grinding and sifting it. The word is from Virginia Algonquian *uskatahomen, usketchamun* 'that which is treated' (in a way specified by a preceding element), probably 'that which is ground or beaten'. The term occurs as a placename in several states, but the name of the town **Hominy,** Oklahoma (Osage Co.), is from an Osage personal name, *hą mąnį,* lit. 'night walker'. **Lake Hominy** is nearby.

HOMOLOBI (AZ, Navajo Co.) \huh MOL uh vee, huh MOH luh vee\. Hopi *homol'ovi* 'high mounded place', from *homol* 'to be mounded'. Also spelled **Homolovi.**

HONAN (AZ, Coconino Co.) From Hopi *honani* 'badger'.

HONOBIA (OK, Le Flore Co.) \HOH nuh buh, huh NOH buh\. Said to be named for a Choctaw named *O-no-bi-a.* Nearby is the **Honobia Creek Wildlife Management Area.**

HONTUBBY (OK, Le Flore Co.) \HON tub ee, hon TUB ee\. Probably the war name of a Choctaw man, perhaps *ǫtabi* 'arrived and killed', from *ona* 'to arrive' and *abi* 'to kill'.

HOONAWPA (AZ, Navajo Co.) Possibly a Hopi name, but its derivation is unclear.

HOOP AND POLE GAME Rock (AZ, Apache Co.) A translation of Navajo *tséyaa nda'azhǫǫshí* 'rock under which the hoop and pole game is played', containing *nda'azhǫǫsh* 'hoop and pole game'.

HOPI \HOH pee\. The name of this Pueblo people, of the Uto-Aztecan language family, is also the name of their reservation in northern Arizona. The term represents their own word *hopi*, meaning 'well mannered, civilized'. There is evidence that, at an earlier period, the Hopis called themselves *mookwi*, and this term entered Spanish in the sixteenth century in the form *moqui*, pronounced \mokwi\. By the nineteenth century however, the word had been reinterpreted in Spanish as \moki\ and had entered English as **Moqui**, pronounced \MOH kee\. The latter appears in placenames such as **Moqui Spring** (AZ, Coconino Co.). The Hopi people found this pronunciation offensive because of its similarity to their word *mooki* 'dies, is dead', and the name was officially changed to *Hopi*. The tribal capital is at **KYKOTSMOVI** (Navajo Co.).

HORSE ROCK (AZ, Apache Co.) Corresponds to Navajo *tsé łį́į́lí* 'horse rock', with *tsé* 'rock'.

HOSKININNI Mesa (AZ, Navajo Co.) \HOS ki nee nee\. Named for a Navajo leader who died in 1909. His Native name was *hashké neiniihí* 'he distributed them with angry insistence', referring to Hoskinini's giving sheep to returning Navajos after the U.S. government forcibly confined them at Bosque Redondo, near Fort Sumner, New Mexico, from 1864 to 1868.

HOSPAH (NM, McKinley Co.) \HOH spah\. From Navajo *haasbá* 'gray area slopes upward'.

HOSTA Butte (NM, McKinley Co.) \HOH stuh\. Named for a nineteenth-century leader of Jemez Pueblo (Tanoan), also known as Francisco; his full Native name is said to have been *Waash e hoste* 'the lightning'.

HOSTEEN BEGAY Well (AZ, Apache Co.) \hos teen buh GAY\. Named for a Navajo man called *hastiin bighe'*, lit. 'man's son', from *hastiin* 'man' and *bighe'* 'his son'. The word *hastiin* is, as in this case, used as an equivalent of English 'mister', and *bighe'* 'Begay' is used as a family name.

HOSTEEN TSO Canyon (AZ, Apache Co.) \hos TEEN soh\. Named for a Navajo man called *hastiin-tsoh* 'big man', containing *hastiin* 'man, Mister'.

HOTASON VO (AZ, Pima Co.) \hoh tuh son VOH\. From Tohono O'odham (Papago) *hodai ṣon wo'o* 'rock-spring pond', combining *hodai* 'rock', *ṣon* 'spring', and *wo'o* 'pond'.

HOTAUTA Canyon (AZ, Coconino Co.) \hoh TOU tuh\. Named for a leader of the Havasupai people. The placename is also spelled **Hotouta**.

HOTEVILLA (AZ, Navajo Co.) \HOH tuh vil uh, hot uh VIL uh, HOHT vil uh\. In Hopi, *hotvela*, from *hó'atvela* 'juniper slope' (*hohu* 'juniper'; *atvela* 'slope'), which this location was called before it was established as a village. The spelling **Hotevila** is also used.

HOT NA NA Wash (AZ, Coconino Co.) \haht NAH nah\. From Navajo *ha'naa na'ní'á* 'bridge across', from *ha'naa* 'across' and *na'ní'á* 'bridge'.

HOTULKEE (OK, Pottawatomie Co.) \huh TOH kee\. Named for a Muskogee leader called Hotulkee Martha or Edward Bullette. The word *Hotulkee* refers to his membership in the Wind clan, *hotvlkvlke* (related to *hotvlē* 'wind').

HÖWIIPA (AZ, Navajo Co.) Hopi *höwìipa* 'dove spring', from *höwi* 'dove' and *paahu* 'water, spring'.

HOYUBY (OK, Hughes Co.) \hoi UB ee, HOI uh bee\. Named for a Choctaw resident whose name was written *Ho-yw-bbee*. This may have been a war name, 'looked for and killed', from *hoyo* 'to look for' and *abi* 'kill'.

HUACHUCA Peak (AZ, Cochise Co.) \wah CHOO kuh\. This Spanish spelling is said to represent a Tohono O'odham (Papago) name meaning 'it rains here'; Tohono O'odham *ju:k* means 'to rain'. Nearby **Fort Huachuca,** headquarters of the U.S. Army Information Systems Command, has a museum with exhibits on Southwest history and military activities in the area.

HUALAPAI (AZ, Coconino Co.) \WAH luh pie, wah luh PIE\. This placename refers to a tribe of the Yuman family, living in northern Arizona. The name is from Mojave *huuwaalyapay* 'pine tree people', from *huwaaly* 'pine tree'. Related forms are **HAULAPAI** (AZ, Mohave Co.), **Hualpai** (AZ, Yavapai Co.), and **WALAPAI** (AZ, Mojave Co.).

HUISACHE Creek (TX, Jackson Co.) \wee SAH chee\. Flows into Lavaca Bay; the name is from Mexican Spanish *huisache* 'mimosa', from Aztec *huixachin.*

HUK OVI (AZ, Navajo Co.) \huk OH vee\. Hopi *huk'ovi* 'high windy place', from *huukyangw, huk-* 'wind'.

HULAH (OK, Osage Co.) \HYOO luh, HOO luh\. Perhaps from Osage *xüðá* 'eagle'. Nearby are **Hulah Lake** and the **Hulah Wildlife Management Area.**

I

IATAN (TX, Mitchell Co.) \ie TAN\. Named for a leader of the Otoes, a Plains tribe; supposedly so named because of his battles with Comanches, who in the Oto language were sometimes called *Ietan* or *Hietan*.

ILLINOIS (OK) \il i NOI\. A transfer from the state of Illinois and from the Miami (or Illinois) tribe, of the Algonquian language family. The **Illinois River** rises in Arkansas, flows through eastern Oklahoma, and enters the Arkansas River in Sequoyah County, Oklahoma. The **Illinois District** in the historic Cherokee Nation, between 1866 and 1907, had its seat in the town of **Illinois**, named after the river, which is no longer there. The name of the Illinois tribe was recorded in 1725 as *Inoca.* It was early referred to by Europeans as *Ilinoüek* and *Illiniwek,* terms of obscure origin, perhaps related to Miami (or Illinois) *ireniwa* 'man'. The modern form *Illinois* represents a seventeenth-century French spelling, pronounced *ilinwe* at that time. A place called **Illinois Bend** is in Texas (Montague Co.).

INDIAHOMA (OK, Comanche Co.) \in dee HO muh\. A blend of *Indian* and *Oklahoma.*

INDIAN \IN dee uhn, IN din\. The word *Indian* was originally an adaptation of Spanish *indio,* used by early explorers to refer to the inhabitants of *las Indias* 'the Indies'. The term appears in many combinations to form placenames that are said to occur in every state of the United States except Hawai'i. Examples from southwestern states are **Indian Creek** (OK, Major Co.), **Indian Gap** (TX, Hamilton Co.), **Indian Peaks** (NM, Summit Co.), and **Indian Wells** (AZ, Navajo Co.).

INDIAN TERRITORY was the name given after 1830 to the area in eastern Oklahoma that served as the site for relocation of the so-called **FIVE CIVILIZED TRIBES**—Cherokee, Chickasaw, Choctaw, Creek, and Seminole—after they were forcibly uprooted from their original homes in the southeastern states. The history of the area is documented at the Five Civilized Tribes Museum in **MUSKOGEE** (OK, Muskogee Co.) and in the book *And Still the Waters Run: The Betrayal of the Five Civilized Tribes* and other works by Angie Debo.

INDIANADA (AZ, Cochise Co.) A combination of *Indian* with an unidentified element.

INDIANOLA (OK, Pittsburg Co.) A combination of *Indian* with the "euphonic" suffix *-ola,* found in words such as *Crayola* and *granola.* There is also an **Indianola** in Texas (Calhoun Co.).

INDIANAPOLIS Siding (OK, Custer Co.) A transfer from the state of Indiana; a blend of *Indian* with Greek *polis* 'city'.

INDIO (CA, Riverside Co.) \IN dee oh\. From Spanish *indio* 'Indian'. In New Mexico there is also an **Indio Canyon** (Catron Co.) and an **Inditos Draw** (McKinley Co.), the latter from the Spanish diminutive plural, meaning 'little Indians'.

INOLA (OK, Rogers Co.) \ie NOH luh\. In the historic Cherokee Naton; said to be from a Cherokee word meaning 'black fox'.

IOLAND (OK, Ellis Co.) \IE oh land\. Coined from the placename **Iowa**, originally referring to a Plains tribe with headquarters in Perkins, Oklahoma (Payne Co.), plus *land*.

IOWA \IE oh way, IE oh wuh\. This name refers first to the Iowa Indian people, of the Chiwere branch of the Siouan linguistic family. It was then applied to the Iowa River, to the state of Iowa, and to many local features. The origin of the term may be in Santee Dakota (Siouan) *ayúxba* 'sleepy ones'. The **Iowa Tribe of Oklahoma** has its headquarters in Perkins, Oklahoma (Payne Co.). The name of the tribe and the language is sometimes spelled *Ioway*. The placename **Iowa Park** occurs in Texas (Wichita Co.).

ISEEO Tank (OK, Comanche Co.) \ie see OH\. Named for Sergeant I-See-O, a Kiowa scout with the U.S. Army. In Kiowa his name is *áisé̖ó:*, meaning 'very dense smoke'.

ISLETA (NM, Bernalillo Co.) \iz LET uh\. The name applied by whites to this Tiwa-speaking pueblo represents the Spanish for 'little island' (diminutive of *isla* 'island'). The Tiwa name for the place is *šiehwíb-àg* 'flint kick-stick place', referring to pieces of flint kicked along in a race. **YSLETA** in Texas (El Paso Co.) is a variant spelling from the same source.

ITAK (AZ, Pima Co.) Probably from Tohono O'odham (Papago) but of unclear derivation.

ITASCA (TX, Hill Co.) \ie TAS kuh, i TAS kuh\. A transfer name from *Lake Itasca* in Minnesota, considered the headwaters of the Mississippi River. The lake's name was invented by Henry R. Schoolcraft in 1832 and is mock Latin for 'true head(waters)', combining the *-itas* of Latin *veritas* 'truth' and the *ca-* of *caput* 'head'. The original Ojibwa name for the lake is said to have been *omashkooz* 'elk'.

ITTITALAH (OK, Johnston Co.) \it i TAH luh\. In the historic Chickasaw Nation; perhaps from Choctaw *ittitakla* 'between' or from Chickasaw *ittintakla'*.

IYANBITO (NM, McKinley Co.) \i YAHN bi toh\. From Navajo *ayání bito'* 'buffalo's spring', containing *ayáni* 'buffalo, bison'.

J

JACAL DE PALO Spring (NM, Rio Arriba Co.) \huh KAHL duh PAH loh\. The Spanish phrase means 'hut (made) of wood', containing Mexican Spanish *jacal* 'hut', from Aztec *xacalli*.

JACONA (NM, Santa Fe Co.) \hah KOH nuh\. This Spanish spelling corresponds to Tewa *sekǫnæ* 'at the tobacco barranca', from *sa* 'tobacco', *ko* 'barranca', and *-næ* 'place'. The adjacent community of **Jaconita** \hah koh NEE tuh\ is named with the Spanish diminutive.

JACQUEZ Canyon (NM, San Juan Co.) \HAH kuhz\. Named for a Navajo, Candalero Jacquez (or Jaques), who owned grazing allotments there.

JADITO (AZ, Navajo Co.) \JED i toh, juh DIT oh\. From Navajo *jádító* 'antelope spring', combining *jádí* 'antelope' and *tó* 'water'. Alternative spellings are **Jedito** and **Jeddito.**

JAMAICA Beach (TX, Galveston Co.) \juh MAY kuh\. A transfer of the name of the Caribbean island or of the town in New York state (Queens Co.). The New York name is perhaps an abbreviation of a Delaware word meaning 'beaver pond', in Unami Delaware *təmá:kwe* 'beaver'. In any case, the placename was probably modified in English to resemble *Jamaica,* the name of the West Indian island, which is derived from the Taino language.

JEDDITO (AZ, Navajo Co.) \JED i toh\. An alternative spelling of **JADITO.**

JEDITO (AZ, Navajo Co.) \JED i toh\. An alternative spelling of **JADITO.**

JEMEZ Pueblo (NM, Sandoval Co.) \HAY muhs, HAY muhz\. From Towa *hiimiš* 'Jemez people', plural of *hiim* 'Jemez person'.

JHUS Canyon (AZ, Cochise Co.) \joos, hohs\. Named for a nineteenth-century Apache leader; the term reflects the Western Apache name *hosh.*

JICARILLA (NM, Lincoln Co.) \hik uh REE yuh\. Refers to an Apache subgroup, as well as being a placename. The Mexican Spanish word means 'little cup', the diminutive of *jícara* 'drinking cup, originally made of gourd', from Aztec *xicalli.* An alternate Spanish diminutive of *jícara* occurs in **Jicarita Creek.**

JICARITA Creek (NM, Taos Co.) \hik uh REE tuh\. See **JICARILLA.**

JOJOBA Boating Site (AZ, Maricopa Co.) \hoh HOH buh\. *Jojoba* is the name of a desert plant, from Pima *hohowai.* The plant is known as an alternative source for rubber and is used in soaps and cosmetics.

JOKAKI (AZ, Maricopa Co.) \joh KAH kee\. From Hopi *tsöqaki* 'mud house', containing *tsöqa* 'mud'.

JUAN TABO Canyon (NM, Bernalillo Co.) \wahn tuh BOH\. Possibly abbreviated from the name of an eighteenth-century resident named Juan

Taboso. The term *Taboso* referred to an Indian group akin to the Apaches in southeastern New Mexico.

JUANITA BEGAY Spring (AZ, Coconino Co.) \wah NEE tuh buh GAY\. Apparently named after a Navajo woman. Begay is a common Navajo family name; see **HOSTEEN BEGAY Well.**

JUCQA-VA (AZ, Navajo Co.) \juh KWAH vah\. From Hopi *tsöqava*, perhaps 'mud water', from *tsöqa* 'mud'.

JUMANOS, Mesa de los (NM, Torrance Co.) \hoo MAH nohs\. The term refers to a Pueblo group encountered by early Spanish explorers; their language is unknown.

JUNCTION, THE (AZ, Apache Co.) A loose translation of Navajo *ahidiidlíní* 'flowing together'.

JWA QWAW GWA Spring (AZ, Coconino Co.) From Havasupai, of unclear derivation.

K

KACHINA Point (AZ, Apache Co.) \kuh CHEE nuh\. From Hopi *katsina* 'spirit being'. The kachinas are the ancestral spirits represented by dancers in Hopi ceremonies and by the "kachina dolls" made for children or to be sold as souvenirs.

KAHACHI MILIUK (AZ, Pima Co.) On the Tohono O'odham (Papago) Reservation; from Tohono O'odham *ge* 'big', *aji* 'skinny', and *meliwkuḍ* 'place where runners end a race; finish line'.

KAIBAB Plateau (AZ, Mohave Co.) \KIE bab\. An abbreviated form of Southern Paiute *kaipapittsi* 'plateau', from *kaipa* 'mountain'. The **Kaibab National Forest** consists of three districts north and south of Grand Canyon National Park.

KAIBITO (AZ, Coconino Co.) \KIE bee toh\. From Navajo *k'ai' bii' tó* 'willow-inside-spring', containing *k'ai'* 'willow', *bii'* 'inside', and *tó* 'water'.

KAIHON KUG (AZ, PIMA Co.) Tohono O'odham (Papago) *gahon ke:k* 'box standing', containing *kahon* 'box' (from Spanish *cajón*) and *ke:k* 'to be standing'.

KAKA (AZ, Maricopa Co.) On the Tohono O'odham (Papago) Indian Reservation; from Tohono O'odham *gagka* 'a clearing'.

KALHOMA (OK, Pontotoc Co.) \kal HOH muh\. In the traditional Chickasaw Nation; from Chickasaw *kali homma* 'red spring', containing *kali* 'spring, well' and *homma* 'red'.

KANAA Valley (AZ, Coconino Co.) \kuh NAH-ah\. From Hopi *ka'naskatsina* 'name of a kachina'.

KANAWHA (TX, Red River Co.) A transfer name from West Virginia, after an Indian tribe of that area.

KANIMA (OK, Haskell Co.) \kuh NEE muh\. In the traditional Choctaw Nation; from Choctaw *kanihma* 'somewhere'. Perhaps the name was given as a joke.

KANSAS (OK, Delaware Co.) \KAN zus\. Named for the state or for the Indian tribe so called, also called the *Kaw*. The term is a French or English plural of an earlier term, *Kansa*, perhaps derived from a word meaning 'wind'. The term used in the Kaw language for the tribe and its language is *kanza*; spelling variants are *kanze, ká"ze,* and *konze*.

KATEMCY (TX, Mason Co.) \kuh TEM see\. Said to be named for an Indian leader, *Katumse*, language unidentified.

KAVOLIK (AZ, Pima Co.) \KAH vuh lik\. From Tohono O'odham *kawelik* 'hilly, hill'.

KAW City (OK, Kay Co.) The term is an alternative name of the **KANSAS** tribe. Nearby are **Kaw Lake** and the **Kaw Wildlife Management Area.** The headquarters of the **Kaw Nation** is at Kaw City.

KAWAIKAA (AZ, Navajo Co.) \kuh wie KAH-ah\. From the Hopi placename *kawàyka'a,* which is the same as the term for Laguna, a pueblo in New Mexico.

KAY CHEE Draw (NM, McKinley Co.) Perhaps a Navajo personal name containing the surname *Chee,* a shortened form of *łichíí'* 'red'.

KAYENTA (AZ, Navajo Co.) \kay YEN tuh, kah YEN tuh\. From Navajo *tééʼndééh* '(animals) fall into the water'. A related placename is **Tyende Creek** (AZ, Apache Co.).

KECHI (OK, Grady Co.) \KEE chie\. The term refers to an Indian group related to the Caddos and is also spelled **Keechi.** Descendants of the tribe form part of the **Wichita and Affiliated Tribes** at Anadarko, Oklahoma (Caddo Co.). **Keechi Creek**, which flows into the Trinity River in Leon County, Texas, and **Keechie** (TX, Anderson Co.) represent other spellings of the name.

KEETOOWAH. \kuh TOO wuh, ki TOO wuh\. The **United Keetoowah Band of Cherokees,** with its headquarters in Tahlequah, Oklahoma (Cherokee Co.), is a tribe known for conservative Cherokee values. It takes its name from an ancient town on the Tuckasegee River in Swain County, North Carolina, called in Shawnee *kato'hwa.*

KEET SEEL (AZ, Navajo Co.) \keet SEEL\. From Navajo *kits'iilí* 'the place of the shattered houses'. This ancient cliff dwelling, attributed to ancestral Pueblos, is part of **Navajo National Monument**, surrounded by the Navajo Nation.

KEMA (OK, Ottawa Co.) \KEE muh\. Said to be from the Miami (or Illinois) placename *kima.*

KEMAH (TX, Galveston Co.) \KEE mah\. Said to mean 'facing the winds' in an unidentified Indian language.

KENTUCK Creek (TX, Jim Wells Co.) \KEN tuk\. Perhaps an abbreviation of *Kentucky.* It is also possible that the term is a variant of *Kanetuche* or *Kanetuck,* both placenames in Alabama, derived from Choctaw *kantak* 'smilax, a kind of brier from which bread is made'. The term also occurs in **Kentuck Mountain** (AZ, Maricopa Co.).

KENTUCKY Creek (OK, Nowata Co.) \ken TUK ee, kun TUK ee\. Named after the state; the term is perhaps from Iroquoian *gędá'geh* 'at the field', the name of a Shawnee (Algonquian) village in Clark County, Kentucky. Several related transfer placenames in the Southwest are **Kentucky Branch** (TX, Blanco Co.), **Kentucky Gulch** (AZ, Pima Co.), **Kentucky**

Park (a flatland in NM, Colfax Co.), and **Kentucky Town** (TX, Grayson Co.).

KEOKUK Falls (OK, Pottawatomie Co.) \KEE oh kuk\. Named after the town in Iowa or directly after Keokuk, a Sac leader of the early nineteenth century; his Native name was *kiiyohkaka* 'one who moves about alertly'. The placename **Keokuk** also occurs in Texas (Harrison Co.). A related name is **KEOTA**.

KEOTA (OK, Haskell Co.) \kee OH tuh\. Named after a town in Iowa; the Oklahoma site was earlier called *Keoton,* from a blend of *Keokuk* and *Washington.* The transfer name also occurs in New Mexico (Colfax Co.).

KIALEGEE (OK, Hughes Co.) \kie LEE jee\. From the Muskogee tribal town name *kilice.* The **Kialegee Tribal Town of the Creek Indian Nation of Oklahoma** is located at Wetumka.

KIAMICHI River (OK) \kie uh MEE chee, kie uh MEE shee, kie uh MICH ee\. This stream runs into the Red River in Choctaw County. The origin of the name is disputed; perhaps it is a variant of the placename *Kiamesha Lake* in New York state, which is possibly from an Algonquian language and is doubtfully defined as 'clear water'. Perhaps also related is **KIOMATIA** (TX, Red River Co.). Between 1864 and 1907, **Kiamichi County** was part of **PUSHMATAHA** District in the Choctaw Nation. The French word *kamichi,* an allusion to a family of waterfowl, has also been suggested as the source of the name.

KIASHITA (NM, Sandoval Co.) \kee AH shi TAH\. From Jemez (Towa) *k'yaashota* 'place where sheep live', containing *k'yaa* 'sheep'.

KIAWA Mountain (NM, Rio Arriba Co.) \KIE uh wah\. Perhaps a variant of **KIOWA,** the name of a Native American tribe.

KICKAPOO (OK, Caddo Co.) \KIK uh poo\. The name of an Algonquian people that once lived in central Illinois; the Native name is *kiikaapoa,* sometimes interpreted as 'wanderers'. Later on, groups of Kickapoos moved to Kansas, Oklahoma, Texas, and northern Mexico. In Oklahoma, the **Kickapoo Tribe of Oklahoma** has its headquarters at McLoud (Pottawatomie Co.), and **Kickapoo Creek** is in Lincoln County. In Texas, the **Kickapoo Traditional Tribe of Texas** has its headquarters at Eagle Pass (Maverick Co.), **Kickapoo Creek** flows into Lake Lavingston in Polk County, and **Kickapoo Cavern State Park** is in Kinney County.

KIDZIIBAHI (AZ, Apache Co.) Perhaps Navajo for 'gray streak mountain'.

KIIQO (AZ, Navajo Co.) From Hopi *kiiqö* 'ruined house'.

KIM-ME-NI-OLI Ruins (NM, McKinley Co.) \kim uh nie OH lee\. From Navajo *kin bii' naayolí* 'house in which the wind blows around', containing *kin* 'house', *bii'* 'inside', and *naayolí* 'the wind blows about'. Also spelled **Kin Bineola.**

KIN Indian Ruin (NM, San Juan Co.) Coined from Navajo *kin* 'house', a word generally used to refer to Pueblo houses as opposed to Navajo hogans.

KIN HOCHO'I (NM, McKinley Co.) \kin HOHCH kee\. From Navajo *kin hoóchxǫ́ǫ́'í* 'ugly house'.

KIN KLETSO (NM, San Juan Co.) \kin KLET soh\. From Navajo *kin łitsooí* 'yellow house'.

KIN KLIZHIN (NM, San Juan Co.) \kin kli ZHIN, kin KLICH in\. From Navajo *kin łizhiní* 'black house'.

KINLICHEE (AZ, Apache Co.) \KIN li chee, kin LEE chee\. From Navajo *kin dah łichí'í* 'red house up at an elevation'.

KIN NAA DAA (NM, Rio Arriba Co.) \kin nah DAH\. Possibly from Navajo *kin* 'stone house' and *naadą́ą́'* 'corn'.

KIN NAHASBASD (NM, San Juan Co.) From Navajo *kin názbąs* 'circular house'. Also spelled **Kin Nahasbas.**

KIN NAHZIN (NM, McKinley Co.) \kin NAH zin\. From Navajo *kin náázíní* 'house standing erect'.

KIN NIZHONI Ruin (NM, McKinley Co.) \kin ni ZHOH nee\. From Navajo *kin nizhóní* 'beautiful house'.

KIN TEEL (NM, McKinley Co.) \kin TEEL\. From Navajo *kinteel* 'wide house'. A related name is **WIDE RUINS** (AZ, Apache Co.).

KIN YA-AH (NM, McKinley Co.) \kin YAH-ah\. From Navajo *kin yaa'á* 'house that sticks up'. Also spelled **Kin Ya'a.**

KINTA (OK, Haskell Co.) \KIN tuh\. In the traditional Choctaw Nation; from Choctaw *kį́tah* 'beaver'. **Kinta Lake** is nearby.

KIOMATIA (TX, Red River Co.) \kie oh MEE shee, kie uh MAY shee uh\. This name is perhaps a variant of the placename *Kiamesha Lake* in New York state, which is possibly from an Algonquian language and is doubtfully defined as 'clear water'. A related name may be that of the **KIAMICHI River** (OK, Choctaw Co.).

KIOWA \KIE oh wah, KIE oh wuh\. The name refers to a tribe of the southern Great Plains; it appears in early Spanish records as *Caigua* and is spelled *cáuigú* in the current Kiowa writing system. The **Kiowa Tribe** has its headquarters at Carnegie, Oklahoma (Caddo Co.). There is a town called **Kiowa** in Oklahoma (Pittsburg Co.). A stream called **Kiowa Creek** arises in Texas (Lipscomb Co.) and flows into Oklahoma, emptying into the North Canadian River (Harper Co.). The **Kiowa National Grassland** is in Union County, New Mexico. Possibly related is the name **KIAWA Mountain.**

KIVA School (AZ, Maricopa Co.) \KEE vuh\. From Hopi *kiva* 'ceremonial chamber'.

KIWANIS Park (AZ, Maricopa Co.) \ki WAH nis\. Parks of this name throughout the Southwest are named for the Kiwanis Club, a civic organization. The term is from Ojibwa and has been said to mean 'to make oneself known', but in fact it seems to be taken from an entry in an 1880 Ojibwa dictionary: *"Kiwanis (nin).* I make noise; I am foolish and wanton."

KLAGETOH (AZ, Apache Co.) \KLAG uh toh, KLAG toh\. From Navajo *łeeyi'tó* 'water in the ground', containing *tó* 'water'.

KLETHLA Valley (AZ, Coconino Co.) \KLETH luh\. From Navajo *łį́į́' łání* 'horses-many', containing *łį́į́'* 'horse'.

KLONDIKE (TX, Delta Co.) A transfer name from northwestern Canada; from Gwich'in *tr'ondëk* 'hammer-stone stream'.

KO VAYA (AZ, Pima Co.) \koh VAH yuh\. See **COBABI.**

KOHATK (AZ, Pinal Co.) \KOH hatk\. From Pima *kohadk* 'hollow'. There is a **Kohatk Valley** in the same county.

KOHI KUG (AZ, Pima Co.) \koh hee KUK\. On the Tohono O'odham (Papago) Reservation; from Tohono O'odham *gohi ke:k* 'mulberry standing'.

KOKOMO (TX, Eastland Co.) \KOH kuh moh\. Transferred from the city in Indiana. Said to be from Miami (or Illinois) *mahko-kima* 'bear-chief', the name of a tribal leader.

KOKOPNYAMA (AZ, Navajo Co.) \koh kohp NYAH muh\. From Hopi *kookopngyamu* 'the Fire Clan', one of the descent groups into which the Hopi people are divided.

KOM KUG (AZ, Pima Co.) \kohm KUG\. From Tohono O'odham (Papago) *ko:m ke:k,* lit. 'hackberry standing'. The hackberry (*Celtis* spp.) is a tree of the elm family, bearing cherry-like fruit; its Tohono O'odham (Papago) name is the basis of local Spanish *comero* 'hackberry tree'.

KOM VO (AZ, Pima Co.) \kohm VOH\. From Tohono O'odham (Papago) *ko:m wo'o,* lit. 'hackberry pond', containing *ko:m* 'hackberry' and *wo'o* 'pond'.

KOMAK WUACHO (AZ, Pima Co.) From Tohono O'odham (Papago) *komaḑk weco,* lit. 'flats beneath'.

KOMALTY (OK, Kiowa Co.) \koh MAHL tee\. Named for *Ko-mal-te,* a Kiowa leader.

KOMATKE (AZ, Maricopa Co.) \koh MAHT kee\. From Pima *komaḑk* 'flats'.

KOMELIK Mountain (AZ, Pima Co.) \KOH muh lik\. From Tohono O'odham (Papago) *komalĭk* 'low flat place'.

KOMO Point (AZ, Coconino Co.) \KOH moh\. Perhaps the surname of a Havasupai family.

KONAWA (OK, Seminole Co.) \KON uh wah, kuh NAH wah\. In the traditional Seminole Nation; from Muskogee *konawv* 'bead(s)'. **Lake Konawa** is nearby.

KOONKAZACHEY (OK, Kiowa Co.) This settlement was named for a leading Plains Apache chief whose enrolled name was Gonkon but who was known always as "Apache John."

KOSOMA (OK, Pushmataha Co.) \koh SOH muh\. From Choctaw *kosoma* 'stink'.

KOSON VAYA (AZ, Pima Co.) From Tohono O'odham (Papago) *koson wahia*, lit. 'packrat well', from *koson* 'packrat' and *wahia* 'well'.

KOTS KUG (AZ, Pima Co.) \kots KUG\. On the Tohono O'odham (Papago) Reservation; from Tohono O'odham *kots ke:k* 'standing cross', containing *kots* 'cross' (from Spanish *cruz*) and *ke:k* 'stand'.

KOWETA Mission (OK, Wagoner Co.) The first mission school among the Muskogee was established in 1841 at this historic site near the town of **COWETA** in the traditional Creek Nation. The name is from *kvwetv*, a transfer name from a tribal townsite in Georgia (Coweta Co.).

KUAKATCH (AZ, Pima Co.) On the Tohono O'odham (Papago) Reservation; from Tohono O'odham *ku:kaj* 'its end', meaning 'end of the mountain'.

KUAPA Ruin (NM, Sandoval Co.) From a Cochiti Pueblo (Keresan) placename.

KUAUA (NM, Sandoval Co.) \koo AH wah\. From the Tiwa placename *ghowye* 'evergreen'.

KUI CHUT VACHK (AZ, Pima Co.) From Tohono O'odham (Papago) *kui ceḍ wack* 'pond in the mesquite', containing *kui* 'mesquite' and *wack* 'pond'.

KUI TATK (AZ, Pima Co.) \koo ee TAHTK\. On the Tohono O'odham (Papago) Reservation; from Tohono O'odham *kui ṭaṭk* 'mesquite root', containing *kui* 'mesquite' and *ṭaṭk* 'root'. An alternate form is **KUK TATK**.

KUIT VAYA (AZ, Pima Co.) On the (Papago) Reservation; from Tohono O'odham *kui wahia* 'mesquite well', containing *kui* 'mesquite' and *wahia* 'well'.

KUKE CHEDEDAGI Tank (AZ, Pima Co.) From O'odham (Piman) *ku:g cehedagĭ* 'green edges', containing *ku:g* 'edge' and *cehedagĭ* 'green'.

KUK TATK (AZ, Pima Co.) On the Tohono O'odham (Papago) Reservation; from Tohono O'odham *kui tatk* 'mesquite root', containing *kui* 'mesquite' and *ṭaṭk* 'root'. An alternate form is **KUI TATK**.

KULI CHUK CHU (OK, McCurtain Co.) \kul ee CHOOK choo\. In the traditional Choctaw Nation; from Choctaw *kalih* 'well, spring' and *chokcho* 'maple tree'.

KULI INLA (OK, McCurtain Co.) \kul ee IN luh\. In the traditional Choctaw Nation; from Choctaw *kalih* 'well, spring' and *ịla* 'different'.

KULLICHITO (OK, McCurtain Co.) \kul ee CHIT oh\. In the traditional Choctaw Nation; from Choctaw *kalih chito* 'big spring', containing *kalih* 'well, spring' and *chito* 'big'.

KULLITUKLO (OK, McCurtain Co.) \kul ee TOOK loh\. In the traditional Choctaw Naton; from Choctaw *kalih toklo* 'two springs', containing *kalih* 'well, spring' and *toklo* 'two'. In the period 1864–1907, **Kullituklo County** was part of **APUKSHUNNUBBEE District** in the Choctaw Nation.

KULLY CHAHA Township (OK, Le Flore Co.) \kul ee CHAH hah\. In the traditional Choctaw Nation; from Choctaw *kalih chaaha* 'high spring', containing *kalih* 'well, spring' and *chaaha* 'high'.

KUPK (AZ, Pima Co.) \KOOPK\. On the Tohono O'odham (Papago) Reservation; from Tohono O'odham *ku:pĭk* 'dam, dike'.

KUSA (OK, Okmulgee Co.) \KOO suh, KYOO suh\. Although this site is in the traditional Creek Nation, the name is probably from Choctaw *kǫshak* 'cane (the plant)'. The Oklahoma placename is perhaps a transfer from *Coosa,* the name of a Creek town in Alabama.

KWAGUNT Creek (AZ, Coconino Co.) Named for a Southern Paiute resident in the nineteenth century.

KWA'KIN'A Historic Site (NM, McKinley Co.) The site of a Zuni village occupied at the time of European contact in 1540.

KWASTIYUKWA Ruin (NM, Sandoval Co.) \kwah see YOO kwah\. From Jemez (Towa) *kwaan-tiyu-kwa,* lit. 'pine-cicada-place'.

KYKOTSMOVI (AZ, Navajo Co.) \ki KOTS moh vee, kyuh KOTS moh vee, kie kots MOH vee\. From Hopi *kiqö-tsmo-vi* 'place of hills made of ruins' (lit. 'ruins-hill-place'). This place, the Hopi tribal capital, is also known locally as *K-Town*.

KYOTE (TX, Atascosa Co.) \KIE oht\. Probably from the same source as **COYOTE**, a word used in many other placenames and referring to the animal (from Spanish *coyote,* ultimately from Aztec *coyotl*).

L

LADY WHITE Trail (AZ, Apache Co.) A translation of Navajo *asdzą́ą́ łigaaí habitiin* 'Lady White's trail-up-out', named for a Navajo woman.

LAHOMA (OK, Garfield Co.) \luh HOH muh\. A shortening of **OKLAHOMA,** from the Choctaw for 'Indian' (lit. 'people-red').

LATEXO (TX, Houston Co.) \luh TEK soh\. An acronym for the Louisiana and Texas Orchard Company. (See **TEXAS.**)

LAWTONKA, Lake (OK, Comanche Co.) This appears to be an "Indianization" of the name of the nearby city of Lawton.

LECHE-E Wash (AZ, Coconino Co.) \luh CHEE, luh CHAY\. From Navajo *łíchíí* 'red'.

LEFLORE County (OK) \luh FLOR\. In the traditional Choctaw Nation; named for a family of mixed French and Choctaw descent, perhaps from French *Lafleur,* lit. 'the flower'.

LEFT HAND Spring (OK, Blaine Co.) Named for an Arapaho leader who was called "Left Hand" in English; in Arapaho the word is *nowó:θ* 'left side'.

LEMEVA (AZ, Navajo Co.) Probably from Hopi *lemo-va* 'hail-water', containing *lemowa* 'hail'.

LENAPAH (OK, Nowata Co.) \LEN uh pah\. A variant of *Lenape,* referring to the Delaware tribe; the tribe calls itself *ləná:p:e* 'original people'. The **Lenapah-Delaware Census Division** is in this county.

LENNA (OK, McIntosh Co.) \LEN uh\. In the traditional Creek Nation; named for Lenna Moore, a Muskogee resident.

LIMA (OK, Seminole Co.) \LIE muh, LEE muh\. This once mostly African American town is now a ghost town, but there is a **New Lima** near the same location, between **SEMINOLE** and **WEWOKA.** The name comes from the city in Peru, derived through Spanish from Quechua *rimac.* It occurs in several states. The same Peruvian placename of course occurs in *lima beans.*

LIPAN (TX, Hood Co.) \lie PAN, li PAN\. Named for the Lipan Apaches, a tribe originallly living in Texas. In Arizona (Coconino Co.), **Lipan Point** is a famous lookout point over the Grand Canyon.

LIPANTITLAN, Fort (TX, Nueces Co.) \li PAN ti lahn\. This is the historic site of a fort constructed around 1831 by Mexican forces and surrendered to American immigrants in 1835. It was apparently named for a camp of Lipan Apaches in the vicinity, with the Aztec suffix *-titlan* 'place'.

LITTLE (OK, Seminole Co.) Named for Thomas Little, a Seminole leader.

LITTLE ROBE Creek (OK, Dewey Co.) Named for a Cheyenne leader; his Native name was probably *tšeške'oomahe,* lit. 'little-robed-one'.

LITTLE WATER (NM, San Juan Co.) A translation of Navajo *tó-'áłts'íísí*, lit. 'water-little'.

LIZARD Spring (AZ, Apache Co.) A translation of Navajo *na'ashǫ'ii-to'í* 'lizard-spring', containing *na'ashǫ'ii* 'lizard'.

LOHALI Mesa (AZ, Apache Co.) From Navajo *łóó' háátįį* 'fish flow-out', containing *łóó'* 'fish'.

LOKASAKAL Wash (AZ, Navajo Co.) \loh KAH suh kahl\. From Navajo *lók'aa' sikaad* 'reeds spread-out', containing *lók'aa'* 'reeds'.

LOLAMAI Point (AZ, Navajo Co.) \LOH luh mie\. Said to be named for an Oraibi Hopi leader. A related form is *lolma* 'beautiful one'.

LOMAKI Ruin (AZ, Coconino Co.) \loh MAH kee\. From Hopi *loma-ki* 'beautiful house'.

LONG Lake (NM, San Juan Co.) A translation of Navajo *be'ek'id hóneezi* 'lake-long', containing *be'ek'id* 'lake'.

LONG TOM Creek (TX, Polk Co.) Named for a Muskogee Indian leader.

LOTAWATA (OK, McIntosh Co.) \lah duh WAH duh\. Perhaps not directly from a Native American source but a play on the Native name **NOWATA** and a reference to the nearby **Lake EUFALA**, into which **Lotawata Creek** flows. In English, of course, this sounds like "a lot of water," the source of the joke. (There is also a Lotawata, Kentucky.)

LUKACHUKAI (AZ, Apache Co.) \loo kuh CHOO kie, loo kuh CHUK ie\. From Navajo *lók'a'ch'égai* 'reeds extending white', containing *lók'aa'* 'reeds'.

LUKFATA (OK, McCurtain Co.) \luk FAH tuh\. In the traditional Choctaw Nation; the Choctaw term may mean 'white dirt', from *lokfi* 'dirt' and *hata* 'white'.

LULULONGTURKWI (AZ, Navajo Co.) From Hopi *lölöqangw-tuyqa* 'bullsnake point', containing *lölöqangw* 'bullsnake' and *tuyqa* 'outside corner'.

LUMKEE (OK, Hughes Co.) \LUM kee\. In the traditional Creek Nation; the name is perhaps from Muskogee *lvmhe* 'eagle'.

M

McCURTAIN (OK, Haskell Co.) Named for Green McCurtain, a principal chief of the Choctaws. **McCurtain County** (OK) was named for the same family.

McINTOSH County (OK) Named for a prominent Creek family (sometimes spelled MacIntosh), the descendants of Captain William McIntosh, a Scot who married two wives from the Coweat (Lower Creek) people. This family included a number of chiefs.

MAHO, Lake (AZ, Navajo Co.) \MAH hoh\. From Hopi *maho,* a First Mesa personal name.

MAHOGANY Creek (NM, Catron Co.) Thus word was applied to a number of native plants with hard wood resembling that of mahogany, a South American tree. The word entered English in the seventeenth century from an Indian language of the West Indies. In the western United States, it generally refers to a bush called mountain mahogany.

MAHONE Peak (AZ, Yavapai Co.) Named for Jim Mahone, a nineteenth-century Hualapai scout.

MAHTSON-PI (AZ, Navajo Co.) \maht SOHN pee\. From Hopi *matsònpi,* the name of a prehistoric ruin at the foot of Hotevilla mesa. See **HOTEVILLA**.

MAIITOH (AZ, Coconino Co.) From the same source as **MAITO** (AZ, Apache Co.).

MAISH VAYA (AZ, Pima Co.) From Tohono O'odham (Papago) *ma'iş wahia,* lit. 'covered well'.

MAITO (AZ, Apache Co.) \MIE toh\. Navajo for 'coyote spring', from *mą'ii* 'coyote' and *to'* 'water, spring'. A related placename is **Maiitoh** (AZ, Coconino Co.).

MAKGUM HAVOKA (AZ, Pima Co.) From Tohono O'odham (Papago) *makkumĭ hawo'oga* 'pond of the caterpillars', containing *makkumĭ* 'caterpillar(s)' and *hawo'oga* 'their pond'.

MAM-A-GAH (AZ, Pima Co.) From Tohono O'odham (Papago) *mamakai* 'medicine men', plural of *ma:kai* 'medicine man'.

MAMATOTK Mountain (AZ, Pima Co.) Perhaps from Tohono O'odham (Papago) *mamhaḍag* 'branch' or *mamhaḍog* 'algae'.

MANAKACHA Point (AZ, Coconino Co.) Named for a leader of the Havasupai people in the early twenthieth century; the Native name was *manakaja.*

MANGAS Mountain (AZ, Gila Co.) \MANG gus\. Probably for Mangas Colorado ("Red Sleeves"), a prominent nineteenth-century Apache leader; the expected Spanish form would be *Mangas Coloradas.* The placename **Mangas** also occurs in New Mexico (Catron Co.).

MANITOU (OK, Tillman Co.) \MAN i too\. This represents a word that occurs in many of the Algonquian languages and means 'spirit, deity, supernatural being'. It first entered English as a borrowing from Unami Delaware *monə́t:u* and was then reinforced by borrowings from other Algonquian languages (in Munsee Delaware *manútoow*, in Ojibwa *manidoo*).

MANKO (OK, Pottawatomie Co.) \MANG koh\. Said to be named for *Me-an-ko,* a Potawatomi resident.

MANUELITO (NM, McKinley Co.) \man wuh LEE toh\. Named for a Navajo leader of the nineteenth century; the Spanish name means 'little Manuel'. His Native name was *ch'il haajiní* 'a black patch of weeds extends out and up', containing *ch'il* 'plant, weeds'.

MANY ARROWS (NM, McKinley Co.) A translation of Navajo *k'aa łaní,* lit. 'arrow(s) many'.

MANY CHERRY Canyon (AZ, Apache Co.) Corresponds to Navajo *didzé łání nástłah* 'many berries cove'.

MANY FARMS (AZ, Apache Co.) Corresponds to Navajo *dá'ák'eh haláni* 'many fields or farms', from *dá'ák'eh* 'field, farm' and *haláni* 'many (over an area)'.

MANY WATERS (AZ, Apache Co.) Corresponds to Navajo *tó łání* 'many springs', from *tó* 'water' and *łání* 'much, many'.

MARAMEC (OK, Pawnee Co.) From the Miami (or Illinois) word for 'catfish'; related forms are Meskwaki *mya:name:kwa* and Ojibwa *maanameg.*

MARIANO Lake (NM, McKinley Co.) \mah ree AY noh\. Named for a Navajo leader of the late nineteenth century. The Native placename was *be'ek'id hóteelí* 'broad lake', containing *be'ek'id* 'lake' and *teel* 'broad'.

MARICOPA County (AZ) \mare i KOH puh\. This name refers to a people of the Yuman linguistic family; it is short for Cocomaricopa, a term of unknown origin, applied by the Spanish around 1690. The present-day Maricopas call themselves *piipáa* 'people'.

MASIPA Spring (AZ, Navajo Co.) \muh SEE puh\. From Hopi *masìipa* 'gray spring', containing *masìi* 'gray'.

MASSACRE (NM, Sierra Co.) \MAS uh ker\. This term has been applied in many states to places where fighting between Indians and whites or between Indian groups produced many dead.

MASSAI Point (AZ, Cochise Co.) \muh SIE, MAS ee\. Named for an Apache warrior known in English as *Big-foot Massai;* his Native name is unknown.

MATKATAMIBA Canyon (AZ, Coconino Co.) Named for a Hualapai family.

MATOAKA (OK, Washington Co.) \muh TOH uh kuh\. This was supposedly the private name of Pocahontas, said to be derived from Virginia Algonquian *metawake* 'she amuses herself playing with something'.

MAYES County (OK) In the traditional Cherokee Nation; named for Samuel H. Mayes, a Cherokee leader.

MAYTUBBY Springs (OK, Atoka Co.) Maytubby is a Chickasaw surname. Floyd E. Maytubby become governor of the Chickasaws in 1939.

MAZATZAL Peak (AZ, Gila Co.) \mah zuh TSAHL, mad uh ZEL\. The name is said to be Apache for 'bleak, barren', but the derivation is unclear.

MEAT ROCK (AZ, Apache Co.) A translation of Navajo *tsé-'atsį'* 'meat rock'.

MECOSUKEY (OK, Seminole Co.) \mek uh SOO kee\. The term refers to a Muskogean people and their language, now spoken among the Seminoles of southern Florida. Also spelled *Mikasuki,* it refers to an old tribal town in southern Georgia, perhaps derived from *miki* 'chief'. The alternative spelling **Mekusukey** is also used in Oklahoma (Seminole Co.).

MEDICINE. This word has been widely used in Indian English and in discussions of Indian cultures to refer not only to plant products used for curing diseases but also to anything to do with ceremonial practices or supernatural power; it could thus be translated in many contexts as 'sacred'. The term occurs frequently in placenames, as in **Medicine Creek** (OK, Comanche Co.) and **Medicine Mound** (TX, Hardeman Co.). The term *medicine lodge* can refer to a structure such as a sweathouse in which worship and healing were performed; an example of its use as a placename is **Medicine Lodge River** (OK, Alfalfa Co.).

MEKKO (OK, Pittsburg Co.) Probably from Muskogee *mēkko* 'chief'. The alternative spelling **MICCO** also occurs in Oklahoma (McIntosh Co.).

MEKUSUKEY (OK, Seminole Co.) \mek uh SOO kee\. See **MECOSUKEY.**

MENDOTA (TX, Hemphill Co.) \men DOH tuh\. A transfer name from Minnesota; derived from Dakota *mdóte* 'mouth of a river, junction of one river with another'.

MERIWHITICA Canyon (AZ, Mohave Co.) From a Havasupai placename, *mat widita,* lit. 'earth hard'.

MESCAL (AZ, Cochise Co.) \mes KAL\. This Mexican Spanish term is from Aztec *mexcalli* and designates the fleshy, edible parts of several desert plants, including species of agave and yucca. Indians used it chiefly for food but also to make rope, baskets, and other products. Today it is also used to make distilled liquor. As a place name, **Mescal** is widespread in the Southwest—for example, in New Mexico (Sierra Co.) and Texas (Bandera Co.).

MESCALERO (AZ, Coconino Co.) \mes kuh LARE oh\. This placename refers to an Apachean people of the Southwest and is Spanish for 'maker of mescal'. The placename also occurs in New Mexico (Otero Co.), where the Mescalero Apache Reservation is located.

MESQUITAL Tank (NM, Quay Co.) \mes kee TAHL\. This is a Mexican Spanish term meaning 'mesquite grove', from *mesquite* 'mesquite tree', from Aztec *mizquitl*.

MESQUITE (AZ, Pima Co.) \mes KEET\. The term refers to a desert tree bearing edible beans; it is borrowed from Aztec *mizquitl* by way of Spanish. As a placename, **Mesquite** is common in the Southwest—for example, in Oklahoma (Jackson Co.) and Texas (Dallas Co.).

METATE (AZ, Graham Co.) \muh TAH tee\. Mexican Spanish for 'grinding slab', from Aztec *metlatl*. The placename also occurs in Texas (Atascosa Co.).

MEXHOMA (OK, Cimarron Co.) \meks HOH muh\. This name was made up from elements of *New Mexico* and *Oklahoma*, because it is near the state boundary.

MEXICAN Springs (NM, McKinley Co.) The term corresponds to Navajo *naakaii bi-to'*, lit. 'Mexicans their-spring', from *naakaii* 'Spaniard, Mexican' and *to'* 'spring, water'.

MEXICO. The name of the nation and of its capital city is from Aztec *mexihco*, the name of the ancient capital. Besides occurring in the name of the state of **New Mexico,** related placenames are **Mexico City** (NM, Lea Co.) and **Mexico Creek** (OK, Beaver Co.).

MIAMI \mie AM ee\. The name refers to the Miami (or Illinois), an Algonquian tribe, once living in Ohio, who called themselves *myaamiwa* 'downstream person'. In Oklahoma, the Miami have their headquarters at the town of **Miami** \mie AM uh\, the county seat of Ottawa County. There is also a **Miami** in Arizona (Gila Co.). The Oklahoma placename **Miami** may represent a blend of *Mima*, a woman's name, and the Ohio placename.

MICCO (OK, McIntosh Co.) \MEE koh\. In the traditional Creek Nation; from Muskogee *mēkko* 'chief'. An alternative spelling is **MEKKO** (OK, Pittsburgh Co.).

MICHIGAN (AZ, Apache Co.) \MISH i guhn\. A transfer from the name of the state. It is perhaps from an Ojibwa word *meshi-gami* 'big lake'; consider modern *michaa* 'be big' and *-gami* 'lake'.

MILPAS, Cañada de las (NM, Sandoval Co.) \MIL pus\. A Mexican Spanish name meaning 'creek of the cornfields'. The word *milpa* 'cornfield' is from Aztec *milpan* 'in the field', containing *milli* 'field'.

MILPITAS Draw (NM, McKinley Co.) \mil PEE tus\. The Mexican Spanish name means 'little cornfields', a diminutive plural form of *milpa* 'cornfield'.

MIMBREÑO Point (AZ, Coconino Co.) \mim BRANE yoh\. Named for an Apache group; the word *mimbreño* means 'having to do with willow twigs', from *mimbre* 'willow twigs, withes'.

MIMSH WAHIA Spring (AZ, Pima Co.) \mimsh WAH hee ah\. Perhaps from Tohono O'odham (Papago) *mimṣ wahia* 'Protestants' well', from *mimṣ* 'Protestants' (plural of *mi:ṣ* 'Protestant') and *wahia* 'a well'. The word for 'a Protestant' is *mi:ṣ,* rather surprisingly derived from Spanish *misa* 'Catholic mass'.

MINCO (OK, Grady Co.) \MING koh\. In the historic Chickasaw Nation; corresponds to Chickasaw *minko'* or Choctaw *mįko,* meaning 'chief'.

MINEOLA (TX, Wood Co.) A transfer name from **Mineola** in New York state (Nassau Co.), perhaps containing an Algonquian root, *min-* 'good'.

MINGO (OK, Tulsa Co.) \MING goh\. From Choctaw *mįko* 'chief'. The placename **Mingo** in Texas (Denton Co.) may have the same origin, or it may derive from a personal name used in the novels of James Fenimore Cooper, based on Delaware *ménkwew* 'treacherous'.

MINNEHAHA The name of the heroine of Longfellow's *Hiawatha* has been used as a placename in many states. In 1849, the name for a waterfall near Fort Snelling, Minnesota, was reported as *Mine-hah-hah,* said to be Dakota for 'laughing waters'. In 1855, Longfellow used the name in his poem. A town named **Minnehaha** is in Arizona (Yavapai Co.), and **Minnehaha Creek** is in Oklahoma (Blaine Co.).

MISHONGNOVI (AZ, Navajo Co.) \mi SHONG goh vee, mish HONG goh vee\. From the Hopi placename *musangnuvi.*

MISSOURI City (TX, Fort Bend Co.) \mi ZOOR uh\. This name, along with the name of the Missouria tribe, a Siouan Indian group, and the names of the Missouri River and the state of Missouri, comes from a Miami (or Illinois) (Algonquian) word, *mihsoori* 'big boat'. The modern **Otoe-Missouri Tribe** has its headquarters at Red Rock, Oklahoma (Noble Co.).

MOCCASIN Creek (TX, Gillespie Co.) This term for Indian footwear was originally adopted into English from Virginia Algonquian *mockasin.* It was also adopted to refer to a type of snake, the water moccasin, and to a fish and a flower. The use of *moccasin* in placenames may refer to any of these meanings. In Oklahoma, **Moccasin Bend** is a turn in the Spring River in Ottawa County.

MODOC Valley (OK, Ottawa Co.) \MOH dok\. This placename is the name of an Indian group closely associated with the Klamath people, living on the Oregon-California border. The word is from Klamath *mo:wat'a:kkni:* 'southerners'. The group's name became well known because of the Modoc War of 1872–73, in which Captain Jack led 57 Modocs to hold off as many as 1,000 American soldiers armed with howitzers. They killed

45 soldiers, including General Canby, the only U.S. general to lose his life in an Indian war. After the Modoc War, Captain Jack was executed, and some Modocs were resettled in Oklahoma. The **Modoc Tribe of Oklahoma** now has its headquarters in Miami, Oklahoma (Ottawa Co.).

MOENAVE (AZ, Coconino Co.) \moh NAH vee\. From Hopi *mawyavi* 'where they pick things'.

MOENKOPI (AZ, Coconino Co.) \moh EN koh pee, MOHN koh pee, moh uhn kop ee\. From Hopi *mùnqapi* 'place of flowing water'.

MOGOLLON Mountains (NM, Catron Co.) Named for a former Native people of the precontact Southwest. There are also a **Mogollon** (Catron Co.) and a **Mogollon Box Campground** (NM, Grant Co.).

MOHAVE County (AZ) \muh HAH vee, muh HAV ee\. Named for the **Mojave** Indian tribe, whose home was on both the California and Arizona sides of the Colorado River. The term is now officially spelled with a *j*, both in regard to the people and with reference to placenames in California. The word is derived from the Mojaves' name for themselves, *hamaxáav.* **Lake Mohave**, a part of Lake Mead Recreation Area, extends down the Colorado River 67 miles from Hoover Dam.

MOHAWK \MOH hahk, MOH hawk\. This name refers to a people living in upper New York State and adjacent areas of Canada and to their language, which is of the Iroquoian linguistic stock,. The word was first used in English in 1638 as *Mohowawogs* (plural), derived from a southern New England Algonquian word for the Mohawk people. There is a town called **Mohawk** in Oklahoma (Tulsa Co.) and a **Camp Mohawk** in Texas (Brazoria Co.).

MOIVAVI (AZ, Maricopa Co.) From Tohono O'odham (Papago), of unclear derivation, possibly meaning 'much water' or 'many wells'.

MOKAAC Wash (AZ, Mohave Co.) The name may be from Southern Paiute, but the derivation is unclear.

MONTAZONA Pass (AZ, Pima Co.) \mon tuh ZOH nuh\. A blend of the names *Arizona* and *Montana.*

MONTEZUMA (AZ, Yavapai Co.) \mon tuh ZOO muh\. This is a version of the name of the Aztec ruler at the time the first Spaniards arrived in Mexico. The usual Spanish form is *Moctezuma*; the original Aztec form was *motecuzoma,* meaning something like 'angry lord'.

MOQUI (AZ, Coconino Co.) \MOH kee\. This term was formerly used in both Spanish and English to refer to the Hopi people, but it was dropped because of a confusion of words. The Hopis are said at one time to have called themselves *mookwi,* which Spaniards transcribed as *moqui.* This transcription, however, was also used for the Hopi word *mookii* 'to die', and the pronunciation \MOH kee\ was regarded as offensive when applied to the people. It was replaced by the Native term *hopi* 'peaceful, polite, civilized'.

MOQUITCH Canyon (AZ, Coconino Co.) \MOH kwich\. From Southern Paiute *moːkwic* 'Hopi', itself borrowed from Hopi *mookwi*.

MORI Mesa (AZ, Coconino Co.) \MOR ee\. Said to have been named for a Navajo resident, Hosteen Mori. The name does not sound Navajo; it may be from Hopi *mori* 'bean'.

MOSHULATUBBEE District (OK) Constituted the northern part of the Choctaw Nation between 1864 and 1907. The name is that of a Choctaw leader in the early nineteenth century; it is a war name, originally *amosholi-t-abi* 'he who tries hard and kills', from *amosholi* 'to try hard' and *abi* 'to kill'.

MOTS'OVI (AZ, Navajo Co.) \moh TSOH vee\. From Hopi *mots'ovi* 'high yucca place', combining *mots* 'yucca', *'o* 'high', and *vi* 'place'.

MOWA'API (AZ, Navajo Co.) \moh WAH pee\. From Hopi *mowàapi* 'moist place', combining *mowàa* 'moist' and *pi* 'place'.

MUAV Canyon (AZ, Coconino Co.) \MOO ahv\. From Southern Paiute *mïa-* 'a pass, a divide'.

MUIK VAYA (AZ, Pima Co.) From Tohono O'odham (Papago) *moik wahiam,* lit. 'soft well'.

MULGULLO Point (AZ, Coconino Co.) Named for a Havasupai family.

MUNSEY (OK, Texas Co.) \MUN see\. This name is an alternative spelling of *Munsee,* the name of a language and people, a branch of the Delaware Indians. The Native term is *mǝn'siːw,* referring to a person from *Minisink* (NY, Orange Co.); this in turn is from *mǝ(nǝ́sǝnk,* probably an archaic word for 'on the island'; modern Munsee *mu(náhan* means 'island'. There is also a **Munsy** in Texas (Floyd Co.).

MUSCOGEE Nation. See **MUSKOGEE.**

MUSHAWAY Peak (TX, Borden Co.) \MUSH uh way\. The name may be from Comanche, but the derivation is unclear. *Mushaway* is reported from Texas as a family name.

MUSKINGUM (TX, Ector Co.) \musk ING um\. Transferred from Burlington County, New Jersey; perhaps originally from Shawnee *mǝshkeekwaamǝchki* 'where the land is swampy' or from Delaware *mooskinkum* 'elk's eye'.

MUSKOGEE \musk OH gee\. This is the name of a major Indian group of the southeastern United States, also spelled **Muscogee** and also called **Creek,** belonging to the Muskogean language family. They formerly lived mostly in Georgia and Alabama; now they are in the Muscogee (Creek) Nation of Oklahoma and in the Seminole nations of Oklahoma and Florida. The Native name is *maskoké* or *mvskoké.* The **Muscogee (Creek) Nation of**

Oklahoma has its capital at Okmulgee (Okmulgee Co.), and **Muskogee County,** Oklahoma, has its seat at **Muskogee.**

MUSKRAT Hollow (OK, Delaware Co.) The term refers to an aquatic rodent and comes from *musquash,* from Abenaki *moskwas.* By folk etymology it was changed to include the English word *rat.*

MUTTELOKE Church (OK, Creek Co.) \mut i LOH kee\. In the traditional Creek Nation; this is a Creek congregation near the town of Briscoe. The name is from the Muskogee placename *mvtelokke.*

N

NAABA ANI (NM, San Juan Co.) \nah buh AH nee\. From Navajo *naabi'ááni* 'the place of the enemy's cave', containing *naa'* 'enemy' and *bi'áan* 'his cave'.

NA AH TEE (AZ, Navajo Co.) \nah AY tee\. From Navajo *náá'á dįįh* 'toadstool, locoweed' (lit. 'eyes become none').

NAASHOIBITO (NM, San Juan Co.) \nah SHOI bi toh\. From Navajo *náshdóí bito'* 'wildcat's spring', from *náshdóí* 'wildcat' and *bito'* 'its spring' (*bi-* 'his, hers, its', *-to'* 'water').

NAAT'A'ANIL BIKIN (NM, McKinley Co.) \nah TAH nee bi kin\. This may be Navajo *naat'áanii bikin* 'chieftain's or leader's house', from *naat'áanii* 'chief, leader' and *bikin* 'his house' (*bi-* 'his', *kin* 'house').

NACLINA (TX, Nacogdoches Co.) \nak LEE nuh\. A blend of elements from the names **NACOGDOCHES** County and *Angelina* County.

NACO (AZ, Cochise Co.) \NAH koh, NAY koh\. A shortening of the placename **Nacosari** in nearby Sonora, Mexico; perhaps derived from a word for 'prickly pear' in Opata, a language of Sonora.

NACOGDOCHES (TX, Nacogdoches Co.) \nak uh DOH chis\. This name is the plural of *Nacogdoche,* the French rendering of the name of a division of the Caddo people. **Lake Nacogdoches** is also in the county of the same name.

NAGEEZI (NM, San Juan Co.) \nah GEE zee\. From Navajo *naayízí* 'a squash' (vegetable), lit. 'the one that rolls around'.

NAGGIA (AZ, Pima Co.) \NAHG yah\. From Tohono O'odham (Papago) *naggia* 'hanging'.

NAGOOLTEE Peak (NM, Eddy Co.) \nah GOOL tee\. Probably from Western Apache *naagołtį* 'rain'.

NAKAI YAZZIE Spring (AZ, Navajo Co.) \nuh kie YAH zee, nah ki YAH zee\. From Navajo *nakaai yázhí,* lit. 'Mexican small', containing *nakaai* 'Spaniard, Mexican' and *yázhí* 'small'; probably a personal name. A related placename is **Nokai Canyon** (AZ, Navajo Co.).

NALAKIHU Ruin (AZ, Coconino Co.) \NAL uh kyoo\. From Hopi *nalakihu* 'lone house', containing *nala, naala* 'alone' and *kihu, kiihu* 'house'.

NAMBÉ (NM, Santa Fe Co.) \nam BAY\. From Tewa *nqmbe'e* 'roundish earth'.

NANIH WAIYA, Lake (OK, Pushmataha Co.) From Choctaw *nanih wayya,* lit. 'hill leaning'; this is also the name of a place in Mississippi (Winston Co.).

NANKOWEAP Mesa (AZ, Coconino Co.) \nang KOH weep, NANG kuh weep\. Perhaps from Southern Paiute *nakukkwippa* 'a battle' or *nakukkwi-uippi* 'battle canyon'.

NARBONA Pass (NM, San Juan Co.) \nahr BON uh\. Named for Narbona, a Navajo leader. The pass was given this name, replacing the earlier "Washington Pass," when it was realized in 1992 that the earlier name had commemorated not President George Washington but rather Colonel John Washington, whose soldiers had scalped the Indian leader at this location in 1849. Ironically, Narbona had been a proponent for peace. The personal name *Narbona* was derived from Spanish; his Navajo name was *hastiin naat'áanii,* lit. 'man chief'. The Navajo name for the pass is *béésh łichíi'ii bigiizh* 'copper gap', from *béésh łichíí* ' copper' (*béésh* 'flint, metal' and *łichíí'* 'it is red'), *bi-* 'its', and *-giizh* 'cut, gap, pass'.

NASCHITTI (NM, San Juan Co.) \NAHS chit ee\. From Navajo *nahashch'idí* 'badger', lit. 'the one who digs or scratches around', from *nahashch'id* 'to dig around, scratch around'.

NASHBITO (AZ, Apache Co.) \NASH bi toh, nash BEE toh\. From Navajo *na'shǫ'ii bito'* 'lizard's spring', containing *na'shǫ'ii* 'lizard', *bi-* 'its', and *to* 'spring'.

NASHOBA County (OK, Pushmataha Co.) \nuh SHOH buh\. In the traditional Choctaw Nation; from Choctaw *nashobah* 'wolf'. In the period 1864–1907, Nashoba County was part of **APUKSHUNNUBBEE District** in the Choctaw Nation; its county seat was at **ALIKCHI.**

NATANES Creek (AZ, Graham Co.) \nuh TAN us, nat uh NEEZ\. Perhaps a Spanish plural from Western Apache *nadą́* 'corn'.

NATANNI NEZ School (NM, San Juan Co.) \nuh tah nee NEZ\. From Navajo *naat'áanii nééz* 'tall chief', containing *naat'áanii* 'chief, leader' and *nééz* 'tall'. The name refers to William T. Shelton, school superintendent at Shiprock starting in 1903.

NATCHES (AZ, Graham Co.) \NACH uz\. The name of a Chiricahua Apache leader. As a placename, it is also written **Natchi** (AZ, Coconino Co.). Not to be confused with *Natchez,* a tribal name and placename from the southeastern United States.

NATCHEZ Nation (OK, Sequoyah Co.) \NAH chee, NACH ee\. This is the name of an Indian community living at Notchietown, near Gore, Oklahoma. It represents descendants of a tribe that earlier lived on the lower Mississippi River.

NAVAHO. An alternative spelling of the name of the **NAVAJO** tribe, no longer in general use.

NAVAJO \NAH vuh hoh, NAV uh hoh\. A people and language of the Athabaskan linguistic family. The **Navajo Indian Reservation,** in Arizona, New Mexico, and Utah, is the largest American Indian nation in the United States, with its capital at Window Rock, New Mexico. The

English term *Navajo* is from Spanish *Navajó,* used in the seventeenth century for northwestern New Mexico and said to mean 'large planted fields' in an Indian language. The source is probably Tewa *navahu:,* from *nava* 'field' and *hu:* 'valley'. In 1969 the Navajo Nation officially adopted the spelling *Navajo,* instead of *Navaho.* As a placename, **Navajo** is the name of a county in Arizona and a town in New Mexico (Rio Arriba Co.). **Navajo National Monument** (AZ, Navajo and Coconino Cos.) is a group of precontact cave pueblos surrounded by the Navajo Nation. The **Navajo Mountains** in Oklahoma (Jackson Co.) represent a transfer name. There is a **Navajo Lake State Park** in New Mexico (San Juan Co.).

NAVASOTA (TX, Grimes Co.) \NA vuh SOH tuh\. Said to be from an Indian word, *Nabatoto* 'muddy waters', in an unidentified language. The **Navasota River** is nearby.

NAWT VAYA (AZ, Pima Co.) \naht VAH yuh, nawt VAH yuh\. From Tohono O'odham (Papago) *noḍ wahia* 'pampas-grass well', containing *noḍ* 'pampas-grass' and *wahia* 'well'.

NAZLINI (AZ, Apache Co.) \nahz LEE nee\. From Navajo *názlíní* 'where the water makes a turn as it flows', containing *názlį́* 'to flow in a circle, to turn flowing'.

NECHES (TX, Anderson Co.) \NECH is\. Named for the **Neches River**, which enters the Gulf of Mexico in Jefferson County and was named for a nearby Indian tribe in the Caddo Hasinai confederacy in 1869. The placename **Weches** (Houston Co.) is said to have been adapted from the name of the river.

NEHAWASKI Spring (OK, Osage Co.) Possibly of Osage origin.

NELAGONEY (OK, Osage Co.) \nee LAG uh nee, nee LAH guh nee\. A post office existed here from December 11, 1906, to February 28, 1959. The name is from earlier Osage *ni ðákðį,* lit. 'water good'; the modern form is *ni ðá:lį.*

NEOSHO River (OK) \nee OH shoh\. This major stream runs into the Arkansas River in Cherokee County. The name is from Osage *nį óžo* 'principal river', from *nį* 'water, river' and *óžo* 'principal'.

NESUFTANGA (AZ, Navajo Co.) From Hopi *naqsöptanga* 'deafness medicine', from *naqsöp, naqsövi* 'deaf', *ta* 'causative', and *nga(hu)* 'root, medicine'.

NEW MEXICO The name of this state has its source in the name of Mexico the country, which comes from the Aztec *mexihco,* the name of the Aztecs' ancient capital.

NEWALLA (OK, Oklahoma Co.) \nuh WAW luh, niu WAH luh\. Represents the Osage name for the Canadian River, from *ni* 'water, river' and *bráða* 'wide'.

NEZ Spring (AZ, Coconino Co.) \nez\. From Navajo -*neez* 'long, tall', also used in English as a Navajo family name.

NITSIN Canyon (AZ, Coconino Co.) This Navajo name is said to refer to an antelope drive, but the derivation is unclear.

NIZHONI Point (AZ, Apache Co.) \ni ZHOH nee\. From Navajo *nizhóní* 'it is beautiful'. The placename also occurs in New Mexico (Santa Fe Co.).

NOBONNI DAHNA'A (NM, Cibola Co.) This Zuni name may contain *noponni* 'face' and *tahna* 'to be sticking out'.

NOCONA (TX, Montague Co.) \noh KOH nuh\. Named for a Comanche man, Pete Nocona, whose name was also written *Nookoni*. The word is from *noohkoni*, the name of a Comanche band, meaning 'wandering around', lit. 'moving camp around in circles', from *noo-* 'to haul' and *kooni* 'to turn around'.

NOIPA KAM (AZ, Pima Co.) From the Tohono O'odham (Papago) placename *nowipakam*.

NOKAI Canyon (AZ, Navajo Co.) \nah KIE, nuh KIE, NOH kie\. From Navajo *naakai* 'Spaniard(s), Mexican(s)', lit. 'wanderers'. Navajos also use the word as a family name. A related placename is **NAKAI YAZZIE Spring** (AZ, Navajo Co.).

NOKAITO Bench (AZ, Apache Co.; UT, San Juan Co.) From Navajo *naakaii tó*, lit. 'Mexican water, Mexican spring', containing *tó* 'water'.

NOLIC (AZ, Pima Co.) \NOH lik\. From Tohono O'odham (Papago) *nolik* 'the bend'.

NOPAL (TX, Presidio Co.) \NOH pal\. This Mexican Spanish word refers to the prickly pear cactus, from Aztec *nopalli*.

NOPALERAS Creek (TX, Jim Wells Co.) \noh puh LARE us\. From Spanish, meaning 'little patches of prickly pear cactus', from *nopal* 'prickly pear' (ultimately from Aztec *nopalli*).

NOT TAK TANK (AZ, Pima Co.) Probably from Tohono O'odham (Papago) *noḍ* 'prickly poppy' and *tak* 'sitting' or *tatk* 'root'.

NOTCHIETOWN (OK, Sequoyah Co.) \NOCH ee toun\. Named for Indians of the Natchez tribe, who originally lived in the lower Mississippi River area.

NOTTAWA (TX, Wharton Co.) Probably a transfer name from Michigan (St. Joseph Co.). The term is from an Algonquian word, perhaps Ojibwa *naadwe*, referring to neighboring Iroquoian and Siouan peoples.

NOWATA (OK, seat of Nowata Co.) \NOH wah tuh, noh WAH tuh\. Said to be a Unami (Delaware) word meaning 'welcome'; the original form may be *no:wi:t:i* 'come here, little one' or *nuwi ta* 'come here'. An alternative source sometimes cited, also from Delaware, is *nuwatu* 'I know something'.

NUMU Creek (OK, Comanche Co.) From Comanche *nʉmʉ* 'Comanche, people'. A renaming of the former Squaw Creek. (See **SQUAW Mountain.**)

NUNNA (OK, McCurtain Co.) \NUN uh\. In the traditional Choctaw Nation; from Choctaw *nona* 'cooked'.

NUYAKA (OK, Okmulgee Co.) \noo YAH kuh\. In the traditional Creek Nation; from Muskogee *nuyakv*. The name represents a transfer of the name *New Yorker* from Georgia (Heard Co.), the site of a Muskogee village, perhaps so named after a delegation of Muskogee leaders visited New York to negotiate a treaty in 1790. The nearby **Nuyaka Creek** flows into the Deep Fork River.

O

OAK Ridge (AZ, Apache Co.) Corresponds to Navajo *tséch'il yílk'id* 'oak hill', from *tséch'il* 'oak' (*tsé* 'rock', *ch'il* 'plant') and *yiílk'id* 'hill'.

OBI Point (AZ, Coconino Co.) \OH bee\. Probably from Southern Paiute *opi* 'wood, trees'.

OCAW (TX, McLennan Co.) \oh KOU\. A backward spelling of **Waco**, from the name of a Caddoan group.

OCHELATA (OK, Washington Co.) \oh shuh LAY tuh\. In the traditional Cherokee Nation; named for a nineteenth-century Cherokee leader whose English name was Charles Thompson.

OCONEE (OK, Coal Co.) \oh KOH nee\. From the Creek word *okoni eknoni*, 'great big water', a transfer from a historic Creek town on the Chattahoochee River in Georgia. Also related to the Muskogee placename written as *Oconi, Ekwoni*, and so forth, said to mean 'place of springs' (*uekiwv* means 'a spring'). Alternatively, the name may be a transfer from **Oconee County,** South Carolina, whose name is said to be derived from that of a Cherokee village, *Ukwûəni*.

OCOTILLO \ok uh TIL oh\. This word is the Spanish name for a brilliantly flowering desert shrub, *Fouquieria splendens*. The Spanish term is the diminutive of *ocote* 'pine, firewood', itself derived from Aztec *ocotl*. **Ocotillo Hills** is in New Mexico (Eddy Co.), and **Ocotillo Siding** is in Texas (Presidio Co.).

OGEECHEE (OK, Ottawa Co.) \oh GEE chee\. Perhaps from Muskogee, meaning 'river of the Uchees', referring to an associated people called **Uchee, EUCHEE,** or **YUCHI;** the Muskogee term is *yocce*.

OHIO (TX, Hamilton Co.) \oh HIE oh\. A transfer name from the Ohio River. From Seneca *ohi:yo'*, a proper name derived from *ohi:yo:h* 'good river'. The Senecas use the term not only for the Ohio River but also for the Allegheny or Allegany River, which they consider to be the headwaters of the Ohio.

OKA ACHUKMA (OK, McCurtain Co.) \OH kuh uh CHUK mah\. In the traditional Choctaw Nation; from Choctaw *oka* 'water' and *achukma* 'good'.

OKARCHE (OK, Canadian Co.) Coined from elements of the words **OKLAHOMA, ARAPAHO,** and **CHEYENNE,** all of Indian origin.

OKEENE (OK, Blaine Co.) \oh KEEN\. Coined from elements of the words **OKLAHOMA, CHEROKEE,** and **CHEYENNE,** all words of Indian origin.

OKEMAH (OK, Okfuskee Co.) \oh KEE muh\. Perhaps named for a Kickapoo chief, or possibly from Ojibwa *ogimaa* 'chief'. **Okemah Lake** is nearby. The placename also occurs in Arizona (Maricopa Co.).

OKESA (OK, Osage Co.) \oh KEE suh\. From Osage *ohkísa* 'half of something, halfway'.

OKFUSKEE County (OK) \ohk FUS kee\. In the traditional Creek Nation; from Muskogee *akfvske,* the name of a tribal town, perhaps meaning 'a point of land between streams'. Named after a historic Creek town in Alabama, whose name in turn refers to the origin of a clan name among the Muskogee people (Creek Indians).

OKLAHOMA \oh kluh HOH muh\. The name of this state was coined by Allen Wright, a Choctaw scholar and Choctaw principal chief from 1866 to 1870. It means 'Indians', from Choctaw *oklah* 'people' and *homma* 'red'. The placenames **Oklahoma City** and **Oklahoma County** are derived from that of the state. There is also an **Oklahoma** in Texas (Montgomery Co.) and an **Oklahoma Mine** in Arizona (Pinal Co.).

OKMULGEE (OK, seat of Okmulgee Co.) \ohk MUL gee\. Perhaps a transfer from **Oakmulgee** in Alabama (Perry Co.). Probably from Hitchiti *oki* 'water' and *molki* 'boiling, bubbling'. The town of Okmulgee was the seat of **Okmulgee District** in the Creek Nation between 1867 and 1907, is the historic capital of the Creek Nation, and is today the site of the headquarters of the Muscogee Creek Nation. Nearby are **Okmulgee Lake** and **Okmulgee Wildlife Management Area.**

OKOEE (OK, Craig Co.) \oh KOH ee\. In the traditional Cherokee Nation; perhaps a transfer from Ocoee, Tennessee (Polk Co.). The original source is the Cherokee placename *uwagahi* 'apricot vine place'.

OKOLONA (OK, McCurtain Co.) In the original Choctaw Nation. Although the source is unclear, the name may come from the Choctaw *oka labali,* 'a place caved in or washed out by water', or perhaps from Choctaw *oklah-lokoli* 'people gathered-together'.

OKTAHA (OK, Muskogee Co.) \ohk TAH hah\. In the original Creek Nation; abbreviated from the name of *Oktaha-sars Harjo,* a Muskogee leader of the nineteenth century (containing *oktahv* 'sand').

OLAGAH A variant spelling of **OOLOGAH.**

OLJATO Wash (AZ, Navajo Co.) \ohl JAY toh\. From Navajo *ooljéé'tó,* from *ooljéé'* 'moon' and *tó* 'water, spring'. An alternative spelling is **Oljeto.**

OLO (AZ, Coconino Co.) \OH loh\. From Havasupai *olo* 'horse', perhaps from Spanish *caballo.*

OLUSTEE (OK, Jackson Co.) \oh LUS tee\. A transfer from Florida (Baker Co.); from Muskogee *ue-lvste* 'black water' (lit. 'water-black').

OMAHA (TX, Morris Co.) \OH muh haw, OH muh hah\. Perhaps transferred from Nebraska, or from the name of a Siouan people and language. The Siouan term is *umą́ hą,* perhaps meaning 'upstream, against the flow'.

ONALASKA (TX, Polk Co.) \on uh LAS kuh\. Modified from the name of *Unalaska,* an island off the coast of Alaska; from Aleut *nawan-alaxsxa,* lit. 'along-there mainland'.

ONAPA (OK, McIntosh Co.) \OH nuh puh\. In the historic Creek Nation; from Muskogee *onvpv* 'above, upper'.

ONEIDA (OK, Kingfisher Co.) \oh NIE duh, oh NEE duh\. This is the name of an Iroquoian people, originally of upper New York state, later moved in part to Wisconsin. In the Oneida language, the people are called *oneyote'a:kâ:* 'people of *one:yóte*'. The term *one:yóte* means 'standing stone', containing *-ney -* 'stone' and *-ot-* 'to stand'.

ONTARIO (OK, Ottawa Co.) \on TARE ee oh\. Transferred from the name of the lake, lying between the United States and Canada; from Mohawk *otarí:io* 'beautiful lake' or perhaps 'great lake', from *-otar-* 'lake, river'.

OOLAGAH (OK, Rogers Co.) \OO luh gah\. An alternate spelling of **OOLOGAH.**

OOLOGAH (OK, Rogers Co.) \OO luh gah\. In the traditional Cherokee Nation; said to have been named for *Oologah,* or Dark Cloud, a Cherokee leader; the name is related to *ulogili* 'cloud'. Near the town are **Oologah Lake, Oologah Wildlife Management Area**, and the **Will Rogers Birthplace Ranch**, childhood home of the famous author and entertainer of Cherokee descent (1879–1935). A variant spelling is **Olagah.**

OOWALA (OK, Rogers Co.) \oo WAH luh\. In the traditional Cherokee Nation; said to be the name of a Cherokee family, known in English as "Lipe."

OPAH (OK, Osage Co.) \OH pah\. From Choctaw *opah* 'owl'.

OPELIKA (TX, Henderson Co.) \oh puh LIE kuh\. Perhaps a transfer name from Lee County, Alabama; from Muskogee *opel-rakko* 'big swamp'.

OPOSSUM Creek (OK, Nowata Co.) \oh POS um\. The name of this marsupial animal is from Virginia Algonquian *opassom,* equivalent to an ancient Algonquian term meaning 'white dog'. There is also an **Opossum Creek** in Texas (Williamson Co.). See also under the abbreviated form **POSSUM.**

ORAIBI (AZ, Navajo Co.) \oh RIE bee, uh RIE bee\. From Hopi *orayvi,* combining *oray,* the name of a rock at the site, and *-vi* 'place'. The village of Old Oraibi, perched high on a rocky mesa, is perhaps the oldest of the present Hopi villages and may be among the oldest continuously inhabited towns of the United States.

OSAGE \OH sayj, oh SAYJ\. A people and language of the lower Missouri River valley. The term is adapted, through French *Osage,* from the Native term *wažáže,* which referred to clan groups in several related languages. The **Osage Nation** has its headquarters at **PAWHUSKA**, Oklahoma (Osage Co.). Nearby are the **Osage Wildlife Management Area** and **Osage Hill State Park**. As a placename, **Osage** also occurs in Texas (Coryell Co.).

OSCEOLA (TX, Hill Co.) \os ee OH luh, oh see OH luh\. The name is that of a Seminole leader who led his people against U.S. forces during the Seminole War, fought in Florida in the early nineteenth century. The Native term is Muskogee *vsse yvholv,* from *vsse* 'leaves used for tea', especially leaves of the yaupon holly (used to prepare a beverage called the 'black drink'), and *yvholv* 'shouter', a term used in war names. Osceola's bravery and military skill made him a national hero.

OSEUMA (OK, Ottawa Co.) \os ee OO muh\. In the traditional Cherokee Nation; said to be from Cherokee *a'siyu ama* 'good water'.

OSOBAVI Peak (AZ, Pima Co.) From O'odham *o:so waw* 'scythe bedrock', containing *o:so* 'scythe' (borrowed from Spanish *hoz*) and *waw* 'bedrock'.

OTEX (OK, Texas Co.) \OH teks\. This name was coined from elements of the placenames **OKLAHOMA** and **TEXAS**, both of Indian origin. Otex is adjacent to **TEXHOMA**.

OTIPOBY Comanche Cemetery (OK, Comanche Co.) \oh ti POH bee\. From Comanche *otɨ-paapi,* a personal name (lit. 'brown water').

OTOE (OK, Noble Co.) \OH toh\. The name of a people and language in the Siouan linguistic family. The term was earlier recorded as *Wah-toh-ta-na,* which may be from *watúhtaną* 'to copulate', or this may be a folk etymology within the Native language. The spelling **Oto** also occurs. The **Otoe-Missouri Tribe** has its headquarters at Red Rock, Oklahoma (Noble Co.).

OTOWI (NM, Santa Fe Co.) \oh TOH wee, OH tuh wee\. Said to be from Tewa, meaning 'gap where water sinks'.

OTTAWA County (OK) \OT uh wuh, OT uh wah\. The name of an Algonquian people, closely related to the Ojibwas, of Michigan and the Canadian province of Ontario. Their Native name is *ota:wa:,* perhaps meaning 'to trade' (related to Ojibwa *adaawe* 'to buy'). The headquarters of the **Ottawa Tribe of Oklahoma** is at Miami, Oklahoma (Ottawa Co.).

OUACHITA (OK, Le Flore Co.) \WOSH i tah, WOSH i taw\. The name is that of a Caddoan people; the term is also familiar from **WASHITA County,** Oklahoma, and from the town of **Wichita** in Kansas. The **Ouachita Wildlife Management Area** is near Hodgen (OK, Le Flore Co.); the **Ouachita National Forest** is in southeastern Oklahoma and adjacent Arkansas, with headquarters in Talihina (OK, Le Flore Co.). The **Ouachita National Recreation Trail** runs from **TALIMENA State Park** (OK, Le Flore Co.) to Pinnacle, Arkansas. An alternative source is *owa chittoh,* Choctaw for 'big hunt', via French.

OZA Butte (AZ, Coconino Co.) \OH zuh\. From Southern Paiute *ottsa* 'water jar', a basket lined with gum to hold water.

OZARK Mountains (OK) \OH zark\. From French *aux Arcs*, short for *aux Arcansas* 'to the Arkansas (people)', meaning to the **QUAPAW** people; this group at one time occupied much of the Arkansas area. The term *Arcansas* (a French plural) was in turn borrowed from Central Algonquian *akansa* 'Quapaw', derived from Siouan *kką:ze,* a term referring to the Dhegiha branch of the Siouan family; this is the source of the name of the *Kansa* people and of the placename **KANSAS.**

P

PA-AKO Ruins (NM, Bernalillo Co.) \puh AH koh\. Perhaps from Tiwa, meaning 'root of the cottonwood'.

PAD WO'O Tank (AZ, Pima Co.) From Tohono O'odham (Papago) *paḍ wo'o* 'evil pond', containing *paḍ* 'bad, evil' and *wo'o* 'pond'.

PADUCAH (TX, Cottle Co.) \puh DOO kuh, puh DYOO kuh\. Perhaps a transfer from Paducah, Kentucky (McCracken Co.); from French *Padouca,* a term referring to a number of peoples in the southern plains, first the Plains Apaches and later the Comanches. The name may come from a Siouan word referring in general to enemy groups—for example, Quapaw *ppattookka,* Osage *ppátǫkka,* and Otoe *phadų́ kha.*

PAGUEKWASH Point (AZ, Mohave Co.) Southern Paiute, meaning 'fish tail', from *pakïu* 'fish' and *kwassipi* 'tail'.

PAHE (OK, Osage Co.) \pah HEE\. Named for an Osage leader called 'Pretty Hair' in English; from *hpahǘ* 'hair'.

PAINT Creek (TX, Concho Co.) Placenames involving *Paint* are likely to refer to places where Indians obtained pigments, especially vermillion, for decoration, or where they used such pigments to make drawings on trees or rocks.

PAIUTE. \PIE yoot\. This name applies to several tribes and languages of the Great Basin. In Arizona, the Southern Paiutes occupied the North Rim of the Grand Canyon and currently live on the **Kaibab Paiute Indian Reservation** on the Arizona-Utah border. The word is not derived from the name of the neighboring Ute tribe in Utah but is from local Spanish *payuchis,* probably borrowed from Southern Paiute *paiyuttsiŋwï* 'those who go and return', containing *paiyu* 'to go and return'. An alternate spelling appears in **Piute** (AZ, Navaho Co.).

PAKOON Springs (AZ, Mohave Co.) From Southern Paiute, perhaps meaning 'bubbling water', containing *paa-* 'water'.

PALUXY (TX, Hood Co.) \puh LUK see\. Perhaps from the same source as *Biloxi,* the name of a Siouan people of the lower Mississippi River, for whom the town of Biloxi, Mississippi is named.

PAMPA (TX, Gray Co.) \PAM puh\. From Quechua, a language spoken in the Andean region of South America, meaning 'a plain'.

PANAMA (OK, Le Flore Co.) \PAN uh mah\. From the name of the Central American nation, said to be an Indian word meaning 'place of many fish'.

PANAMETA Point (AZ, Coconino Co.) Said to be named for a Havasupai man; his Native name has been given as *apá nahamída* 'man feels glad'.

PAN NEPODK (AZ, Pima Co.) Perhaps Tohono O'odham (Papago) for 'shaped like a loaf of bread', from *pa:n* 'bread' (a borrowing from Spanish *pan*) and *ñepodk* 'loaf-shaped'.

PAN TAK (AZ, Pima Co.) On the Tohono O'odham (Papago) Reservation; from Tohono O'odham *ban dak* 'coyote sitting', combining *ban* 'coyote' and *dak* 'to sit'.

PANOLA (OK, Latimer Co.) \pan OH luh, puh NOH luh\. In the traditional Choctaw Nation; the name is derived from Choctaw *ponǫla* or *ponoola* 'cotton'. In the period 1866–1907, a **Panola County** existed in the southeastern part of the Chickasaw Nation. There are also a town called **Panola** and a **Panola County** (with county seat at Carthage) in southeastern Texas.

PANYA Point (AZ, Coconino Co.) From the surname of a Havasupai family.

PAPAGO (AZ, Maricopa Co.) \PAH puh goh, PAP uh goh\. This term refers to an Indian people and language of the Piman branch of the Uto-Aztecan linguistic family; their name is now officially **TOHONO O'ODHAM**, lit. 'desert people'. The local Spanish word *Pápago* is from earlier *Pápabos* and still earlier *Papabi-Ootam*, ultimately from Native *ba:bawĭ-'o'odham* 'people of the tepary beans', containing *ba:bawĭ* 'tepary beans' (sing. *ba:wĭ*) and *o'odham* 'people'. Tepary beans are a wild plant food of the Southwest.

PAPAGUERÍA (AZ, Pima Co.) \pah puh guh REE uh\. The name of this area in southern Arizona, meaning 'Papago country', is from Spanish *Papaguería* 'place of the Papagos', from *pápago* 'Papago'.

PAPALOTE Creek (TX, Refugio Co.) \pah puh LOH tee\. From Mexican Spanish *papalote* 'butterfly, kite, windmill', from Aztec *papalotl* 'butterfly'. The placename also occurs in Arizona (Pima Co.).

PAPOOSE Creek (TX, Hardeman Co.) \pap OOS, puh POOS\. This New England Algonquian word for 'child' was recorded in 1643 as *papoòs* and has been widely used to mean 'Indian child'. At present, however, like the word *squaw*, it is considered offensive.

PARASHANT Canyon (AZ, Mohave Co.) \PARE uh shant, PAHR uh shant\. Probably from Southern Paiute *pa-tasïantï*, lit. 'water-dawn', from *paa-* 'water' and *tasya* 'dawn'. In the northwest corner of Arizona, bordering Grand Canyon National Park, is **Grand Canyon–Parashant National Monument**, an undeveloped area accessible only by unpaved roads. A variant spelling is **Parashont.**

PARIA River (AZ, Coconino Co.) \puh REE uh\. From Southern Paiute *patïkia* 'elk', lit. 'water-deer', from *pa-* 'water' and *tïkia* 'deer'.

PARISSAWAMPITTS Canyon (AZ, Coconino Co.) \pah RUH sing wahm pit si\. From Southern Paiute *patï-sïŋwan-pittsi*, lit. 'sand-gravel-little', from *patïya* 'sand' and *sïŋwa-* 'gravel'. Said to imitate the sound of the spring, bubbles coming out of the ground.

PARMICHO (OK, Coal Co.) In the traditional Choctaw Nation; derived from Choctaw *palah mishsha* 'light in the distance', combining *palah* 'light' and *mishsha* 'far off'.

PASADENA (TX, Harris Co.) \pas uh DEE nuh\. Transferred from California, where one of the founders of the community of that name wrote to a friend who was a missionary to the Ojibwa Indians in Minnesota and asked that he suggest an Indian name for the new California town. The result was *basadinaa* 'valley'.

PATANGVOSTUYQA (AZ, Navajo Co.) Hopi *patangvostuyqa* 'pumpkin seed point', from *patang, patnga* 'pumpkin, squash', *vos, poosi* 'seed', and *tuyqa* 'point'.

PATTOWATOMIE Creek (TX, Henderson Co.) \pah tuh WAH tuh mee\. From the same source as **POTAWATOMI**.

PAUCAUNLA (OK, Bryan Co.) In the traditional Choctaw Nation; perhaps from Choctaw *pakqli* 'to bloom, to flower'.

PAVO KUG Wash (AZ, Pima Co.) Perhaps from Tohono O'odham (Papago) *waw keːg*, lit. 'bedrock-standing', from *waw* 'bedrock' and *keːg* 'to stand'.

PAWHUSKA (OK, Osage Co.) \puh HUS kuh, paw HUS kuh\. Named for an Osage leader of the early nineteenth century; the meaning of the term is 'white hair', from *hpahiiə* 'hair of the head' and *ska* 'white'. **Lake Pawhuska** is nearby.

PAWNEE County (OK) \pah NEE, paw NEE, PAH nee, PAW nee\. The name is that of an Indian people of the central plains and their language, of the Caddoan language family. The Pawnee Indians now live in Oklahoma and call themselves *paári* in their own language; both that name and the English term apparently originated in neighboring Siouan languages— for example, Omaha *ppáðį* and Oto *pányi*. The seat of Pawnee County is the town of **Pawnee**, where a 1910 mansion and ranch buildings of the **Pawnee Bill Ranch Site** house memorabilia of the cowboy, frontiersman, and entertainer Gordon W. Lillie, also known as Pawnee Bill, whose Wild West Show competed with Buffalo Bill's Wild West Show, merging with it in 1908 as the Two Bills' Show. The **Pawnee Nation** has its headquarters in Pawnee, Oklahoma. The placename **Pawnee** also occurs in Texas (Bee Co.).

PAWPAW Creek (OK, Craig Co.) \PAH pah, PAW paw\. This name for a native bush, *Asimina triloba,* and its fruit was borrowed from Spanish *papaya,* referring to a fruit of quite a different kind and probably derived from a Native language of the Caribbean. The placename **Paw Paw Bayou** occurs in Texas (Harrison Co.).

PAYA Point (AZ, Coconino Co.) The surname of a Havasupai family.

PECAN Creek (OK, Craig Co.) \pi KAHN\. The term refers to a kind of tree and its nuts, growing primarily in the southeastern United States. But the word entered English, via French *pacane,* from a generic term for 'nut' that occurs in many Algonquian languages. It may have first entered French from Miami (or Illinois) *pacana, pacane*; the corresponding term in Fox is *paka:ni* 'nut', in Ojibwa *bagaan*, and in Abnaki *bagôn*. The word *pecan* occurs in several Texas placenames, such as **Pecan Bayou** (Red River Co.).

PECOS River (NM and TX) \PAY kohs, PAY kus\. From Spanish *Pecos,* derived from Keresan *p'æyok'ona,* the name of **Pecos Pueblo** (NM, Eddy Co.) as used at Kewa (Santo Domingo) Pueblo. This term is perhaps from the name *p'akyula* in the Towa language, the language spoken by the former residents of Pecos Pueblo, the last remnants of whom moved to Towa-speaking Jemez Pueblo in 1838. In Texas, the name is applied to the town of **Pecos** (Reeves Co.) and to **Pecos County.**

PEJUNKWA Ruin (NM, Sandoval Co.) From Jemez (Towa) *peejunkwa* 'heart hill place', containing *peel* 'heart, soul', *šuun* 'hill', and *-kwa* 'place'.

PELON Creek (TX, Dimmit Co.) \pee LOHN\. The Spanish word *pelón* 'bald' may refer here to Indian groups that shaved their heads; however, the alternative form **PILON** can also mean 'a loaf of sugar' or 'a gratuity'.

PENISTAJA (NM, Sandoval Co.) \pen i STAH hah\. Perhaps a Spanish adaptation of Navajo *bíniishdáhí* 'where I sit leaning against it', containing *binii* 'against it' and *shdá* (*sédá*) 'I sit'.

PEORIA (OK, Ottawa Co.) \pee OR ee uh\. A transfer name from Illinois. The word is the name of a subdivision of the Miami (or Illinois) people, recorded in 1673 as *Peoualen*; the Native name is *peewaareewa, peewaalia.* The headquarters of the **Peoria Indian Tribe of Oklahoma** is in Miami, Oklahoma (Ottawa Co.). The placename **Peoria** also occurs in Texas (Hill Co.).

PEPCHILTK (AZ, Pinal Co.) From O'odham (Pima) *pi:pcul* 'concave'.

PERSIMMON (OK, Woodward Co.) \per SIM un\. The name of the fruit was first recorded from Central Atlantic Coast Algonquian as *pessemmins* and similar spellings, containing ancient Algonquian *-min-* 'fruit, berry'. The placename **Persimmon Gap** occurs in Texas (Brewster Co.).

PESHLAKAI Point (AZ, Coconino Co.) Shortened from Navajo *beesh łikaiitsidí* 'silversmith', containing *beesh łikaii* 'silver' (*beesh* 'flint', *łikaii* 'white'). The related placename **Peshlaki** (AZ, Coconino Co.) reflects the Navajo word for 'silver' by itself.

PIA OIK (AZ, Pima Co.) \pee uh OIK, pee YOY(K)\. On the Tohono O'odham (Papago) Reservation; from Tohono O'odham *pi o'oik* 'not striped'. **Pi Oik** is an alternate spelling.

PIATO VAYA (AZ, Pima Co.) This Tohono O'odham (Papago) name contains *wahia* 'well'.

PICKENS (OK, McCurtain Co.) Named for Edmond Pickens, a prominent Chickasaw.

PICURIS (NM, Taos Co.) \pik yoo REES\. The name of this Tiwa pueblo is Spanish *Picurís,* perhaps from Keres *pee-koo-ree-a* 'those who paint'.

PIESTEWA Peak (AZ, Maricopa Co.) \PEE es ti wah, PIE es ti wah\. Once called **Squaw Peak**, this landmark was renamed Piestewa Peak after Pfc. Lori Piestewa, a Hopi woman who died in the Iraq war in 2003— the first Native American servicewoman ever to die in action.

PILON Creek (TX, Dimmit Co.) \pi LOHN\. Perhaps from Spanish *pelón* 'bald', which may refer to Indian groups that shaved their heads, but the spelling *Pilon* can also mean 'a loaf of sugar' or 'a gratuity'. (See **PELON Creek.**)

PIMA \PEE muh\. The name refers to a Uto-Aztecan people of central Arizona; the term *Piman* includes them along with the closely related **Tohono O'odham** (Papagos), as well as related groups in northwest Mexico. The Native term is probably from *pi ma:c* '(I) don't know', which Spaniards mistook for a tribal name, or from *pima* 'no', as used by southern Piman speakers in Mexico. The tribe now lives near Phoenix (Maricopa Co.) in the **Salt River Pima–Maricopa Indian Community**, on the **Gila River Indian Reservation**, and on the **Maricopa-Ak Chin Indian Reservation.** The Pimas call themselves *Akimel O'odham,* lit. 'river people', in contrast to the Tohono O'odham, lit. 'desert people', formerly called Papagos. A confusing point of terminology is that **Pima County** (AZ) is primarily the home of the Tohono O'odham (Papago) tribe, while the Pima tribe lives mainly in Maricopa County.

PIMERIA Alta (Pima Co.) \pee muh REE uh AHL tuh\. The Spanish term *pimería alta* 'upper Pima country' refers to territory occupied by O'odham speakers in Arizona, as opposed to others in Mexico. (See **PIMA.**).

PINALENO Mountains (AZ, Graham Co.) \pin uh LAY noh\. From Spanish *pinaleño,* referring to an Apachean band; perhaps from Western Apache *bįįh* 'deer'. Alternatively, the term may be Spanish for 'belonging to the pine grove', from *pinal* 'pine grove' (*pino* means 'pine').

PIPYAK (AZ, Pima Co.) Perhaps from Tohono O'odham (Papago), of unclear derivation.

PISINIMO (AZ, Pima Co.) \puh SEE nee moh\. From Tohono O'odham (Papago) *pisinmo'o* 'bison head', containing *pisin* 'bison, buffalo' and *mo'o* 'head'.

PITA (NM, Harding Co.) \PEE tuh\. This American Spanish word means both 'fiber, twine' and 'century plant', from which such fiber is extracted; it may be from a Latin American Indian language.

PITAHAYA Canyon (AZ, Pima Co.) \pee tuh HAH yuh\. This American Spanish word, referring to a type of cactus and its edible fruit, is from the Taino language of the Caribbean.

PITOIKAM (AZ, Pima Co.) \peet wee kuhm\. From Tohono O'odham (Papago), perhaps meaning 'sycamore place'.

PIUTE (AZ, Navaho Co.) \PIE yoot\. See **PAIUTE**.

POCAHONTAS Mountain (OK, Pittsburg Co.) \poh kuh HON tus\. The name is that of daughter of the Virginia Algonquian leader Powhatan; her Native name has been given as *Pokachantesu* 'she is playful'. She was also called **MATOAKA,** with a similar meaning, and that, too, appears as a placename (for example, OK, Washington Co.).

POCASSET (OK, Grady Co.) \poh KAS it\. A transfer name from New England; from New England Algonquian *pok-shau* 'it divides'.

POCOLA (OK, Le Flore Co.) \puh KOH luh\. In the traditional Choctaw Nation; from Choctaw *pokkooli* 'ten'.

POISON IVY Canyon (AZ, Apache Co.) Corresponds to Navajo *íshíshjį́į́zh sikaad nástł'ah* 'poison ivy sands spread out cove', from *íshíshjį́į́zh* 'poison ivy'.

POJOAQUE (NM, Santa Fe Co.) \poh HWA kay\. From Tewa *p'osųŋwǽye* 'drink water place', from *p'o* 'water', *sųŋwǽ* 'to drink', and *-ye* 'down at'.

POLACCA (AZ, Navajo Co.) \poh LAH kuh, poh LAK uh, puh LAK uh\. Named for Tom Polacca, a member of the Hopi-Tewa community.

PONCA City (OK, Kay Co.) \PONG kuh\. This name applies to a people and language in the Siouan linguistic family; in historic times, they have lived in Nebraska and in Oklahoma. The Native name used by the people was *ppą́kk*. The **Ponca Tribe** has its headquarters at White Eagle, just south of Ponca City. **Lake Ponca** is nearby.

PONKA BOK Creek (OK, McCurtain Co.) \PONG kuh bok\. Perhaps Choctaw *pąkki book,* 'grape creek', containing *pąkki* 'grape' and *book* 'creek'.

PONTATOC Canyon (AZ, Pima Co.) \PON tuh tok\. Probably a transfer from **PONTOTOC**.

PONTOTOC (OK, Johnston Co.) \PON tuh tok\. A transfer of the name of Pontotoc County in Mississippi. Earlier written *Punkatuckahly*; probably from Choctaw, meaning 'hanging grape', from *pąkki* 'grape' and *takaali* '(sing.) to hang'. The placename also occurs in Texas (Mason Co.).

POPHERS Creek (TX, Angelina Co.) \POH ferz\. Named for an Indian resident named Popher, Native group unidentified.

PORUM (OK, Muskogee Co.) Named for J. Porum Davis, also called Dave Porum, a Cherokee leader.

POSE'UINGE Archaeological District (NM, Rio Arriba Co.) From Tewa *p'osi'oŋwįkeyi,* lit. 'greenness pueblo-ruin', from *p'osi(wi'i)* 'greenness' and *oŋwįkeyi* 'pueblo ruin' (*oŋwį* 'pueblo', *keyi* 'old').

POSEY (OK, Creek Co.) \POH zee\. Named for Alexander Posey, a nineteenth-century Muskogee poet and journalist; the surname may be from *pose* 'cat', from English *pussy.*

POSSUM \POS um\. A shortening of *opossum,* the name of a native marsupial mammal, which comes from Central Algonqian Coast *opassom,* equivalent to an ancient Algonquian term meaning 'white dog'. **Possum Hollow** is in Oklahoma (Washita Co.), and **Possum Kingdom** is in Texas (Palo Pinto Co.).

POTAPO Creek (OK, Atoka Co.) \poh TAH poh, poh TAP oh\. In the traditional Choctaw Nation; perhaps from Choctaw *pataapo* 'floor, bridge'.

POTATO Hill (OK, Ottawa Co.) \puh TAY toh, puh TAY tuh\. This name for an edible tuber is borrowed from Spanish *batata* 'sweet potato', which was in turn borrowed from the Taino language of the Caribbean.

POTAWATOMI \pot uh WAH tuh mee\. The name of an Algonquian people living around Lake Michigan and, in more recent times, in Kansas and Oklahoma; also spelled *Pottawatomie.* The name of **PATTOWATOMIE Creek** in Texas (Henderson Co.) is from this source. The people's name for themselves is *potewatmi.* A tradition exists that this means 'people of the place of the fire' (in Ojibwa, *boodawe* means 'makes a fire'), but this is likely to be a folk etymology. The **Citizen Potawatomi Nation** has its headquarters at Shawnee, Oklahoma, in **POTTAWATOMIE County.**

POTOSI (TX, Taylor Co.) \poh TOH see, puh TOH see\. From Spanish *Potosí,* originally the Aymara name for a famous silver mine in Bolivia, later transferred to San Luis Potosí in Mexico. The term came to be a metaphor for 'untold riches'.

POTTAWATOMIE County (OK) \pot uh WAH tuh mee\. Named for the **POTAWATOMI** tribe, a nation now headquartered at Shawnee, Oklahoma.

POW-WOW Creek (OK, Roger Mills Co.) \POU wou\. From New England Algonquian *powwaw* 'Indian priest', from ancient Algonquian **pawe:wa* 'he dreams; one who dreams'. The term was adopted into English to mean 'medicine man' and then came to be used in the sense of 'council, conference'.

PUARAY (NM, Sandoval Co.) \POO uh ray\. This name of an ancient Tiwa (Tanoan) village is perhaps from a word meaning 'woman'.

PUEBLITO (NM, Socorro Co.) \pweb LEE toh\. Spanish for 'little pueblo, little village' (the diminutive of *pueblo* 'village, town'), referring to an extension of Ohkay Owingeh (San Juan Pueblo), a Tewa settlement.

PUEBLO \PWEB loh\. The Spanish word for 'village, town' has come to be used to refer to the compact Native settlements in Arizona and New Mexico and to the peoples that occupy them: the Hopi (Uto-Aztecan), Zuni, Keres, and Tanoan peoples; they are thus distinguished from traditionally more nomadic groups of the area, such as the Navajos and Apaches.

PUSHMATAHA County (OK, with county seat at Albion) \poosh muh TAH hah, poosh muh TAW haw\. Named after the **Pushmataha District** of the traditional Choctaw Nation, which was in turn named for Chief Pushmataha of the Choctaws, who fought in the War of 1812 and knew Andrew Jackson. His name was perhaps from *vpushi* 'sprout, sapling' and *imvlhtaha* 'prepared, stored', also a war title. Between 1864 and 1907, the Pushmataha District had its seat at Mayhew. The **Pushmataha Wildlife Management Area** is in the county, near the town of Clayton.

PUTEOSI Canyon (AZ, Coconino Co.) From the surname of a Havasupai family. The spelling **PUTESOY** also occurs.

PUTESOY Canyon (AZ, Coconino Co.) From the surname of a Havasupai family. The spelling **PUTEOSI** also occurs.

PUYE Cliff Dwellings (NM, Rio Arriba Co.) \POO yay\. At Santa Clara Pueblo; from Tewa *pu* 'cottontail rabbit' and *ye* 'to assemble'.

PYOTE (TX, Ward Co.) \PIE oht\. From Spanish *peyote* 'a type of hallucinatory cactus', from Aztec *peyotl*.

Q

QUALACU (NM, Socorro Co.) \KWAH luh soo\. From Spanish *Qualacú* or *Cuelaqu,* originally a Piro placename; perhaps a miscopying of something like *Cualaçú,* for *Cualazú.*

QUANAH (TX, Hardeman Co.) \KWAH nuh\. Named for Quanah Parker, a Comanche leader; his Native name was *kwana* 'smell, be fragrant'. The names **Quanah Mountain** and **Quanah Parker Lake** occur in Oklahoma (Comanche Co.). Quanah Parker's grave is at the Apache Prisoner of War Cemetery near Lawton, Oklahoma (Comanche Co.).

QUAPAW (OK, Ottawa Co.) \KWAH pah, KWAW paw\. An Indian people of the Siouan linguistic family, now living in Oklahoma. The Native name is *okáxpa* 'downstream', a reference to the tribe's being farther south in the Mississippi Valley than other Siouan groups. The headquarters of the **Quapaw Tribe of Oklahoma** is in Quapaw. In an earlier period the Quapaws were called the **Arkansas**. There are streams called **Quapaw Creek** in Oklahoma (Lincoln Co.) and Texas (Harrison Co.).

QUARAI (NM, Torrance Co.) \KWOR ie\. Said to be from Tiwa *kuah-aye* 'bear place'.

QUASSARTE \koh wah SAH tee\. This term is a variant of **COUSHATTA** or **Koasati**, the name of a Muskogean tribe originally living in Alabama; in Muskogee, the term is *kowassate*. The **Alabama-Quassarte Tribal Town** is located at Henryetta, Oklahoma (Okmulgee Co.).

QUEBEC (TX, Presidio Co.) \kuh BEK, kwuh BEK\. The Canadian placename, French *Québec,* is from an Algonquian word for 'straits, narrows'. The exact language origin may be Micmac *gepèq.*

QUELITES (NM, Valencia Co.) \kuh LEE tays\. Spanish *quelite* refers to a kind of edible leafy plant; the word is from Aztec *quilitl.*

QUERECHO Plains (NM, Lea Co.) \kuh RAY choh\. Spaniards gave the name *Querecho* to nomadic peoples they encountered in New Mexico. It is probably from the Pecos equivalent of Jemez *k'yǽlǽcoš* 'Navajos, Apaches', in the Towa language.

QUIBURI Mission (AZ, Cochise Co.) \KEE buh ree\. Earlier written as *Giburi, Guibore,* and so forth; perhaps contains the word *ki* 'houses' from Nevome, a Piman language of northwest Mexico.

QUIGUI (NM, Sandoval Co.) \KEE wee\. From a Spanish spelling, *Quigüi,* from Keresan *dyîiwi,* the name for Kewa (Santo Domingo) Pueblo in some Keresan dialects (other than that of Kewa Pueblo itself, in which the name is *dyîiwa*). A probably related placename is **EL GUIQUE** (NM, Rio Arriba Co.).

QUIHI (TX, Medina Co.) \KWEE hee\. Probably from the same source as **KECHI**, referring to a subtribe of the Caddoan language family. **Quihi Creek** is adjacent.

QUIJOTOA (AZ, Pima Co.) \kee hoh TOH uh, kee hi TOO uh\. From Tohono O'odham (Papago) *giho do'ag,* 'burden-basket mountain'.

QUITAQUE (TX, Briscoe Co.) \KIT uh kwee\. Perhaps from an obscure Indian source, but also perhaps from Mexican Spanish *cuita* 'excrement', from Aztec *cuitlatl.*

QUITO (NM, Taos Co.) \KEE toh\. Perhaps named for the capital city of Ecuador, whose name is originally from Quechua.

QUITOBAQUITO (AZ, Pima Co.) \kee toh buh KEE toh, kwee toh bah KWEE toh\. A Spanish diminutive (with *-ito* 'little') from *Quitovac,* from Tohono O'odham (Papago) *gi'ito wa:k,* containing *wa:k* 'standing water'. See also **QUITOVAK.**

QUITOVAK (AZ, Pima Co.) \kee toh VAHK\. From Tohono O'odham (Papago) *gi'ito wa:k,* containing *wa:k* 'standing water'. See also **QUITOBAQUITO.**

R

RABBIT BRUSH (NM, McKinley Co.) Perhaps corresponds to Navajo *k'iił tsoii* (lit. 'alder which is yellow', from *k'ish, k'iish* 'alder'), a plant called 'rabbit brush' in English.

RABBIT EAR Mountain (NM, Union Co.) Said to have been the personal name of an Indian, of an unidentified ethnic group.

RACCOON (TX, Austin Co.) \rak OON, ruh KOON\. The name of the animal is derived from Virginia Algonquian *aroughcun*. See also **COON.**

RAMONA (OK, Washington Co.) \ruh MOH nuh\. Named after the heroine of Helen Hunt Jackson's influential nineteenth-century romantic novel *Ramona*. Ramona is a Spanish name for a woman, the feminine counterpart of *Ramón* 'Raymond'.

RANEGAS Plain (AZ, Yuma Co.) From Hualapai *hanagas* 'good'.

RAYADO (NM, Colfax Co.) \rah YAH doh\. This Spanish word means 'striped, streaked'; in this placename it may refer to Indians who wore painted lines on their faces.

RED Eagle (OK, Osage Co.) Named for Paul Red Eagle, an Osage leader.

RED Lake (AZ, Coconino Co.) From Navajo *be'ek'id ha-lchíí'*, lit. 'lake area-red'.

RED River (NM, Taos Co.) A translation of the Taos (Tiwa) name for this stream.

REDMOON (OK, Roger Mills Co.) Named for a Cheyenne leader whom whites called Red Moon; his Native name was probably *ese'he ohma'aestse*, from *ese'he* 'moon, sun' and *ohma'aestse* 'that which is red'.

ROANOKE (TX, Denton Co.) \ROH nohk, ROH uh nohk\. A name transferred from Roanoke County, Virginia. Roanoke is also the name of the island that was the site of Sir Walter Raleigh's "lost colony" in North Carolina. Recorded in 1584, the word was perhaps a Virginia Algonquian placename. It may be the same as Virginia Algonquian *rawranock* 'shells used for money', which, in the form *roanoke,* came to be used in that sense in English during colonial times.

ROCK Point (AZ, Apache Co.) A translation of Navajo *tsé łichii' deez'áhí* 'pointed or extended red rock', containing *tsé* 'rock', *łichii'* 'it is red', and *deez'á* 'it is pointed, it extends'.

ROCK EXTENDING INTO WATER (AZ, Apache Co.) A translation of Navajo *tsé táá'á* 'rock extending-into-water'.

ROCK STRUCK BY LIGHTNING (AZ, Apache Co.) A translation of Navajo *tsé bi'oosní'í* 'rock struck by lightning'.

ROCK THAT PEOPLE TURNED INTO (AZ, Apache Co.) The Navajo name, *tsé ná'áz'élí,* has been mistakenly translated as 'rock that they (people) turned into', but it actually means 'rock that water flows around'.

ROCK THEY RAN INTO (AZ, Apache Co.) Corresponds to Navajo *tsé biih nííjahí* 'where people repeatedly run into the rock'.

ROGERS County (OK) In the traditional Cherokee Nation; named for Clem V. Rodgers, father of the Cherokee author and humorist Will Rogers. The town of **Rogers** (OK, Mayes Co.), also in the Cherokee Nation, was named for William C. Rogers, principal chief of the Cherokee Nation from 1903 to 1917.

ROMAN NOSE State Park (OK, Blaine Co.) Named for Henry Caruthers Roman Nose, a Cheyenne leader; his Native name was *vohko'xenehe* or *voo'xenehe,* lit. 'crooked-faced-one'.

ROUGH Rock (AZ, Apache Co.) A translation of Navajo *tséch'ízhí* 'rock that is rough', containing *tsé* 'rock' and *ch'ízh, ch'íízh* 'rough'.

ROUND Rock (AZ, Apache Co.) Perhaps an adaptation of Navajo *tsé nikáni* 'bowl-shaped rock', from *tsé* 'rock' and *niká* 'concave, bowl-shaped'.

S

SAA BETOH (AZ, Navajo Co.) \SAH bi toh\. From Navajo, perhaps representing *tsah bito'* 'water of the needle', from *tsah* 'awl, needle', *bi-* 'its', and *to'* 'water'.

SABANA River (TX, Eastland Co.) Probably from the same source as the name of the nearby town of **SABANNO.** The word may be from *šaawanwa,* the self-designation of the Shawnee tribe, or it might have been adapted from the Savannah River and town in South Carolina and Georgia. The latter name is derived from English *savannah* 'tropical grassland', from Spanish *sabana,* taken from a now extinct language of the West Indies.

SAC \sak\. The term refers to a Central Algonquian people of the Midwest; from the French abbreviation *Saki,* probably from Ojibwa *osa:ki:* 'person of the outlet' (in Ojibwa, *o-* marks ethnic groups). The "outlet" referred to is probably that of the **Saginaw River** in Wisconsin. The name of the Sac for themselves is *asa:ki:waki,* (in their language, usually spelled *Sauk; a-* marks ethnic groups). The Sac people have long been associated, linguistically and culturally, with the Fox group, and the **Sac and Fox Nation** is located at Stroud, Oklahoma (Lincoln Co.). An alternative spelling, usually used for the language, is **Sauk.**

SACAHUISTE Draw (NM, Eddy Co.) \say kuh WEE stay\. This Mexican Spanish name, also written *zacahuiste* and *zacahuistle,* refers to the nolina, a type of yucca; it is from Aztec *zacahuitztli,* lit. 'grass-thorn', from *zacatl* 'grass' and *huitztli* 'thorn'. Related placenames are **ZACAHUISTLE Pasture** (TX, Brooks Co.) and **ZACAWEISTA Ranch** (TX, Wilbarger Co.).

SACATE (NM, Socorro Co.) \sah KAH tay\. This Mexican Spanish word, usually spelled *zacate,* means 'grass, hay'; it is from Aztec *zacatl.* Related placenames are **ZACATA Creek, ZACATES,** and **Arroyo ZACATOSO** (TX, Zapata Co.).

SACATON Flat (NM, Grant Co.) \sah kuh TOHN\. This Mexican Spanish word, usually spelled *zacatón,* refers to a type of fodder grass, lit. 'big grass', from *zacate* 'grass, hay', derived from Aztec *zacatl* 'grass'.

SACATOSA Mesa (NM, San Miguel Co.) \sah kuh TOH zuh\. This Mexican Spanish word, usually spelled *zacatosa,* means 'full of grass', from *zacate* 'grass, hay', derived from Aztec *zacatl* 'grass'. A related placename in Texas is **Arroyo ZACATOSO.**

SAGEEYAH (OK, Rogers Co.) \suh GEE yuh\. In the traditional Cherokee Nation; named for a Cherokee resident, Sageeyah Saunders. The given name is a Cherokee rendering of the Biblical name *Zaccheus.*

SAGINAW (TX, Tarrant Co.) \SAG i naw, SAG i nah\. Perhaps a transfer from Wisconsin. From Ojibwa *sa:ki:na:ŋ* 'in the Sauk country', referring to *osa:ki:* 'Sauk, people of the outlet' (of the Saginaw River). See also SAC.

SAGUARO National Park (AZ, Pima Co.) \suh WAH roh\. From the Mexican Spanish name for a type of cactus, perhaps from the Yaqui language of northwestern Mexico; the modern Yaqui form is *sauwo*. A related form is **Sahuaro** (AZ, Yavapai Co.).

SAHOMA, Lake (OK, Creek Co.) Near the town of Sapulpa; the name is probably a combination of **Sapulpa** and **Oklahoma.**

SAHUARITA (AZ, Pima Co.) \sah wah REE tuh, sah wuh REE tuh\. Originally Spanish *sahuarito* or *saguarito* 'little saguaro', a diminutive of *saguaro*, a type of cactus, with variant spelling **Sahuaro.** See **SAGUARO National Park.**

SAHUARO Tank (AZ, Yavapai Co.) *Sahuaro* is an alternative spelling of **Saguaro.** See **SAGUARO National Park.**

SAKETON (NM, San Juan Co.) \sak uh TOHN\. Represents Mexican Spanish *zacatón* 'a type of grass', from *zacate* 'grass, hay', from Aztec *zacatl* 'grass'.

SALAHKAI (AZ, Apache Co.) From Navajo *tséligai* 'white rock(s)', containing *tsé* 'rock' and *łigai* 'it is white'.

SA'LAKO (AZ, Navajo Co.) This stone pillar was named for the Hopi katsina *Sa'lako,* borrowed from the Zuni *Sha'lako.* A related placename is **SHALAKO Ski Trail** (NM, Taos Co.).

SALINA (AZ, Apache Co.) \suh LEE nuh\. From Spanish *salina* 'salt spring, salt flat', by folk etymology from Navajo *tséláni* 'many rocks', containing *tsé* 'rock' and *łáni* 'many'.

SALT TRAIL Canyon (AZ, Coconino Co.) Corresponds to Navajo *áshįįh nabitiin* 'salt trail', containing *áshįįh* 'salt' and *bitiin* 'trail'.

SAN XAVIER DEL BAC, Mission (AZ, Pima Co.) \san huh VERE del BAK, san huh VERE del BAHK\. This Spanish mission, founded in 1692 by the Jesuit father Eusebio Kino, was named for St. Francis Xavier, with the addition of Tohono O'odham (Papago) *wa:k* 'standing water'. The name follows a pattern common throughout Spanish America, that of a saint's name followed by a local name for the same place. In this case, the saint's name is that of St. Francis Xavier, who apostolized Japan. The area around the mission was established as the San Xavier del Bac Indian Reservation in 1874.

SANDUSKY (TX, Marshall Co.) \SAN dus kee, san DUS kee, sun DUS kee\. Transferred from Ohio. Originally written *Sandouske*; perhaps from Wyandot *sa'ndesti* 'water'.

SANENEHECK Rock (AZ, Navajo Co.) From Navajo *tsé ani'įįhí* 'thief rock', containing *tsé* 'rock' and *ani'įįhí* 'thief'.

SANOSTEE (NM, San Juan Co.) \suh NOS tee\. From Navajo *tsé ałnáozt'i'í* 'where rocks overlap or are layered', containing *tsé* 'rock(s)' and *ałnáozt'i'í* 'they overlap'. The placename also occurs in Arizona (Apache Co.), where it is pronounced \SAN us tee\. See also **TSEANAZTI Creek.**

SANTAN (AZ, Pinal Co.) \SAN tan, san TAN\. A Tohono O'odham (Papago) adaptation of Spanish *Santa Ana* 'St. Anne'.

SAPANA VAYA (AZ, Pima Co.) Perhaps from earlier Tohono O'odham (Papago) *ṣapan wahia* 'soap well', from *ṣapan* 'soap' and *wahia* 'well'. The modern form is *ṣawon wahia*. Both forms of the word for 'soap' are from Spanish *jabón.*

SAPAWE (NM, Rio Arriba Co.) This archaeological site is named for a Pueblo village, perhaps of the Tewa language family.

SAPELLO (NM, San Miguel Co.) \SAH pi yoh\. Perhaps from earlier *Shapellote, Chaptellote*; an early French-Kiowa resident of Taos was named *Chapalote.*

SAPULPA (OK, Creek Co.) \suh PUL puh\. In the traditional Creek Nation; named for a nineteenth-century Muskogee leader, *Sus-pul-ber,* or for his family. The name is perhaps from *isvpvlpakē* 'wrapped with'.

SARATOGA County (TX, Hardin Co.) \sare uh TOH guh\. Transferred from New York state. The name was originally written as *Saraghtoge, Saraghtogo,* and so forth; it is apparently from an Iroquoian language but is of unclear derivation.

SASABE (AZ, Pima Co.) \SAH suh bee, SAS uh bee\. From Tohono O'odham (Papago) *ṣaṣawk* 'echo'. The Buenos Aires National Wildlife Refuge is located here, protecting the endangered masked bobwhite quail and other grassland wildlife.

SASAKWA (OK, Seminole Co.) \suh SAH kwuh\. In the traditional Seminole Nation; from Muskogee *sasakwv* 'goose'.

SASE NASKET (AZ, Coconino Co.) Said to be from Navajo, but of unclear derivation.

SASETANHA Mesa (AZ, Apache Co.) Also written *Sastanha*; perhaps from Navajo, but the derivation is unclear.

SAUK. An alternative spelling of **SAC**, usually used in reference to the language of the Sac people.

SAVANNA (OK, Pittsburg Co.) Perhaps a transfer from Georgia. The placename is derived from *Savana*, a name applied to Indians in the Georgia area. Some writers have held that the term is derived from *šaawanwa,* the self-designation of the **Shawnee** Indians, an Algonquian group, but others have argued that the tribal designation is in fact derived from the English topographical term *savannah* 'a type of tropical grassland', which was borrowed by English from Spanish *sabana* in the 1500s and has been used in the southeastern United States since that time. The Spanish term was in turn borrowed in the West Indies from Taino *zabana*. **See also SABANA River.**

SAWIK Mountain (AZ, Maricopa Co.) From Tohono O'odham (Papago) *ṣawikuḍ* 'rattle'.

SAWOKLA (OK, Muskogee Co.) \suh WOH kluh\. In the traditional Creek Nation; from Muskogee *svwokle*, the name of a tribal town.

SAYDATOH (NM, Sandoval Co.) \SAY duh toh\. Perhaps from Navajo, containing *tó* 'water, spring' or *tooh* 'body of water'.

SCHUCHK (AZ, Pima Co.) \s'CHOOCH'k\. From Tohono O'odham (Papago) *scuck* 'black ones', refering to black hills, from *s-* 'adjective', *cu-* 'reduplication', and *cuk* 'black'. *Reduplication* means repetition of a syllable, in this case to form a plural; the second *u* is contracted out.

SCHUCHULI (AZ, Pima Co.) \s'choo CHOO li\. From Tohono O'odham (Papago) *s-cuculig* 'many chickens', from *cucul* 'chicken'.

SCHUK COWLIK (AZ, Pima Co.) \s'chook KOW lik\. From Tohono O'odham (Papago) *scuk kawlik* 'black hill', containing *scuk* 'black' and *kawlik* 'hill'.

SCHUK TO'AG (AZ, Pima Co.) \s'chook toh AHG\. From Tohono O'odham (Papago) *scuk to'ag* 'black mountain'.

SCHUNYAK (AZ, Pima Co.) \s'choon YAHK\. Said to be from Tohono O'odham (Papago), meaning 'much corn', but of unclear derivation.

SCRAPER (OK, Cherokee Co.) \SKRAY pur\. Named for Captain Archibald Scraper, a Union Army officer of Cherokee ancestry during the Civil War.

SEAMA (NM, Cibola Co.) \say AH muh\. From Laguna *ts'íyâama* 'doorway, passageway' or, by extension, 'mountain gap'.

SEBA DALKAI (AZ, Navajo Co.) \say buh DAHL kie\. From Navajo *séí bídaagai* 'sand piled against the ridge', containing *séí* 'sand'.

SEBOYETA (NM, Cibola Co.) \seb oh YET uh\. From Spanish *cebolleta* 'wild onion', a diminutive of *cebolla* 'onion'. The name corresponds to Navajo *tł'ohchin* 'onion', from *tł'oh* 'grass'; a related word is *halchin* 'to have an odor'. The spelling **Cebolleta** is also used.

SEELOCU (NM, Socorro Co.) The name of a Piro pueblo, abandoned during the period of Spanish rule.

SEGEKE Butte (AZ, Navajo Co.) Said to be Navajo for 'square rock', containing *tsé* 'rock'.

SEGETOA Spring (AZ, Apache Co.) An alternate spelling for **Tsegito Spring**, from Navajo *tsiyi' tóhí* 'spring in the forest', containing *tsiyi'* 'forest' and *to* 'water'.

SEGI Mesas (AZ, Navajo Co.) \SEG ee\. From Navajo *tséyí'* 'large canyon', lit. 'inside the rocks', from *tsé* 'rock' and *yi'* 'inside'. The placename is also written **Tsegi**, as in **TSEGI Canyon** (AZ, Navajo Co.**).**

SEGIHATSOSI Canyon (AZ, Navajo Co.) \seg ee huh TSOH see\. From Navajo *tséyi' hats'ózí* 'canyon which is narrow', containing *tséyí'* 'canyon', *ha-* 'area', and *ts'óz* 'narrow'. Alternative spellings are **Tse Ye Ha Tsosi** and **Tseye Ha Tsosi.**

SEI BE TOH Well (AZ, Coconino Co.) \SAY bi toh\. From Navajo *séí bii' tó* 'spring in the sand', containing *séí* 'sand', *bii'* 'in it', and *tó* 'water'.

SEI DEELZHA (AZ, Apache Co.) \say DALE zhah\. From Navajo *séí deelzha* 'jagged sand dune', containing *séí* 'sand' and *deelzhah* 'undulating line'.

SEI HAASGAII Wash (AZ, Coconino Co.) \say HAHS gie\. From Navajo *séí haasgai* 'sand slopes upward white', containing *séí* 'sand' and *haasgai* 'it slopes upward white,' from *lgai* 'to be white'.

SEMINOLE County (OK) \SEM i nohl\. The term refers to a tribe speaking a language of the Muskogean language family, originally living in southern Alabama and Georgia, who moved to Florida during the eighteenth and nineteenth centuries to escape European domination. In Florida they joined escaped African American slaves to form a new tribe. The derivation of the word is from Muskogee *semvnole*, from *semvlone* 'untamed', from Spanish *cimarrón* 'untamed'. Subsequently, a portion of the Florida Seminoles moved to Oklahoma. At present, some of the Florida Seminoles use the Hitchiti (or Mikasuki) language, and others use a variety of Muskogee (Creek). The Oklahoma Seminoles use their own variety of Muskogee; their historic capital is at **WEWOKA** (Seminole Co.) As a placename, **Seminole** occurs in Oklahoma (Seminole Co.) and in Texas (Gaines Co.) **Seminole Canyon State Historic Site,** in Texas (Val Verde Co.), is the site of cave artwork by prehistoric Indians, though the name Seminole is misleading.

SENECA (NM, Union Co.) \SEN uh kuh\. This is the name of an Indian people of the Iroquoian family, living in upper New York state. The term was earlier written as *Sinnekens, Senakees,* and so forth, and was also applied to the Oneidas (another Iroquoian group). It has been proposed that *Seneca* is from a Mahican word meaning 'people of the place of the stone', but this has not been confirmed. As a placename, Seneca occurs not only in New York state (Seneca Co.) but also as a transfer name in many other states—probably with influence from the name of Seneca, the ancient Roman philosopher and dramatist. In New Mexico (Union Co.), however, the placename **Seneca** is said to be a distortion of Spanish *ciénega* 'marsh, meadow'. The headquarters of the **Seneca-Cayuga Tribe of Oklahoma** is in **MIAMI**, Oklahoma (Ottawa Co.).

SENECU (NM, Socorro Co.) The site of a Piro pueblo, destroyed in the Spanish colonial period; the placename was also written as *Tze-no-que* and *She-an-ghua.*

SEQUO (OK, Sequoyah Co.) An abbreviation of **SEQUOYAH**.

SEQUOYAH County (OK) \see KWOI yuh\. In the historic Cherokee Nation; named for a Cherokee man known in English as George Guess or George Gist, who was born around 1776 in eastern Tennessee. His Cherokee name has been explained as *ssiquaya* 'pig's foot', presumably referring to a physical deformity, or, alternatively, as a word, approximately *sigwoyá*, meaning 'one who does more'. It has also been spelled *Sequoia* in English and came to be used as the name for the giant redwood trees of California. Around 1809, he invented a syllabic writing system for his language. **Sequoyah's Home Site**, near Sallisaw, Oklahoma (Sequoyah Co.), is where Sequoyah built a cabin in 1829, after his resettlement to Oklahoma. The **Sequoyah District** of the Cherokee Nation had its seat, between 1851 and 1907, at the former town of **Sequoyah**; this district then became the modern Sequoyah County. In Oklahoma there is also a modern town of **Sequoyah** (Rogers Co.) and the **Sequoyah National Wildlife Refuge**, near Vian (Sequoyah Co.), bordering the Arkansas River. **Sequoyah Bay State Park** is in Wagoner County, and **Sequoyah State Park** is in Cherokee County. A movement existed just before Oklahoma's statehood to form a state named Sequoyah in Indian Territory, in the eastern half of what is now Oklahoma. The popular vote was in favor of the state of Sequoyah, but the U.S. Congress did not allow it.

SETSILTSO Springs (AZ, Apache Co.) \set SILT soh\. From Navajo for 'big oak', from *tséch'il, chéch'il* 'oak' (*tsé, ché* 'rock', *ch'il* 'plant') and -*tsoh* 'big'.

SEVEN LAKES (NM, McKinley Co.) Corresponds to Navajo *tsosts'id be'ek'id* 'seven lakes', from *tsosts'id* 'seven' and *be'ek'id* 'lake'.

SHABIKESHCHEE (NM, San Juan Co.) \shuh BIK uh shee\. From Navajo *shá bik'e'eshchį́* 'the sun is written or carved upon it', containing *shá* 'sun' and *bik'e'eshchį́* 'it is carved on it' (*bi-* 'its', *k'e, k'i* 'upon', *eshchį́* 'it is written or carved').

SHAKAVEYOK (AZ, Pima Co.) \shah kuh vee YOHK\. From Tohono O'odham (Papago) *skakawiyugk* 'many horses', containing *kawiyu* 'horse', from Spanish *caballo*.

SHALAKO Ski Trail (NM, Taos Co.) \SHAL uh koh\. From Zuni *Sha'lako,* the name of a katsina. A related placename is **Sa'lako** (AZ, Navajo Co.).

SHANUB Point (AZ, Mohave Co.) \shuh NUB\. From a Southern Paiute family name, derived from *sinna'api* 'dog, wolf'.

SHAOTKAM (AZ, Pima Co.) \shaht KAHM\. From Tohono O'odham (Papago) *ṣa:ḍkam* 'wild sweet-potato'.

SHAVANO Park (TX, Bexar Co.) \SHAV uh noh\. Probably transferred from Colorado. Named for a Ute leader, whose name, however, seems to be related to *shawano*, an Algonquian term meaning 'southerner'.

SHAWNEE \shah NEE, SHAH nee, shaw NEE, SHAW nee\. This is the name of an Algonquian people, probably once concentrated in southern Ohio but later moved to areas such as Pennsylvania, Kentucky, Alabama, Kansas, and Oklahoma. They were referred to by the French in 1648 as *Ouchaouanag,* corresponding to their self-designation, *šaawanwa,* lit. 'person of the south'. Currently the tribe lives in three major areas: the so-called Loyal Shawnees, formerly regarded as part of the Cherokee Nation, have their headquarters in Jay, Oklahoma (Delaware Co.); the "Absentee Shawnee" band has its headquarters in Shawnee, Oklahoma; and the "Eastern Shawnee of Oklahoma" tribe is in fact not in Oklahoma but in Seneca, Missouri. There are towns called **Shawnee** in Oklahoma (Pottawatomie Co.) and Texas (Angelina Co.).

SHENANDOAH (TX, Montgomery Co.) \shen un DOH uh\. From the Oneida family name *Skenondoah* or *Skenandore,* representing *skęnǫtǫ́ha',* shortened from *oskęnǫtǫ́ha',* derived from *oskęnǫtǫ́ha'* 'deer'. The name was apparently transferred to the Shenandoah River and Valley in Virginia, and from there to many other states.

SHINARUMP Cliffs (AZ, Coconino Co.) \SHIN uh rump\. Probably from Southern Paiute *sïnna'a-rïmpi* 'dog rock', from *sïnna'a-* 'wolf, dog, coyote' and *tïmpi* 'rock'.

SHINUMO Creek (AZ, Coconino Co.) Said to be from a Southern Paiute term referring to 'ancient people, cliff dwellers'.

SHIPAULOVI (AZ, Navajo Co.) \shi POU luh vee, shi POL uh vee\. Also written **Shipolovi** and **Sipaulovi.** This is from the Hopi placename *supawlavi,* containing *-vi* 'place'; the translation 'mosquito-place', which has often been given, cannot be confirmed.

SHIVWITS Plateau (AZ, Mohave Co.) \SHIV wits\. From the name of a Southern Paiute band, *sipici-mï,* perhaps originally meaning 'easterners'.

SHONTO (AZ, Navajo Co.) \SHON toh, SHOHN toh\. From Navajo *shą́ą́'tóhí* 'where the spring is in the sunshine', containing *shą́ą́'* 'in the sunshine' and *tó* 'water'.

SHOOTING ARROWS IN THE ROCK (AZ, Apache Co.) A translation of Navajo *tsé né'eltoho* 'shooting arrows into the rock' (contains *tsé* 'rock'), referring to archery competitions.

SHUNGOPAVI (AZ, Navajo Co.) \shung GOP uh vee, shung GOH puh vee, shi MOH puh vee\. From Hopi *songòopavi* 'sand-grass spring', from *songòo, songowu* 'sand grass, a type of reed', *pa, paahu* 'water, spring', and *vi* 'place'.

SHUSH BE TOU (AZ, Apache Co.) \SHUSH bi toh, SHOOSH bit oh\. From Navajo *shash bito'* 'bear's spring', containing *shash* 'a bear' and *bito'* 'its water' (*bi-* 'its', *to'* 'water, spring').

SIA (NM, Sandoval Co.) \SEE uh\. A variant writing of **ZIA**, the name of a Keresan pueblo in the Rio Grande Valley.

SICHOMOVI (AZ, Navajo Co.) \si CHOM uh vee\. From Hopi *sitsom'ovi* 'flower-hill high-place', combining *si-*, *sihu* 'flower', *-tsom-*, *tsomo* 'hill', *-'o-* 'high', and *-vi* 'place'.

SIF VAYA (AZ, Pinal Co.) From Tohono O'odham (Papago) *siw wahia* 'bitter well', containing *siw* 'bitter' and *wahia* 'well'.

SIKORT CHUAPO (AZ, Pima Co.) \SEE kurt chee uh puh\. From Tohono O'odham (Papago) *sikol ce:po*, lit. 'round bedrock-mortar'.

SIKUL HIMATK (AZ, Pima Co.) From Tohono O'odham (Papago) *sikol himadek* 'whirling (of water)', from *sikol* 'round' and *himadk* 'going'.

SIKYATKI Ruins (AZ, Navajo Co.) \sik YAHT kee\. From Hopi *sikyatki*, containing *sikya* 'ravine' and *tk* 'cut'.

SIL MURK (AZ, Maricopa Co.) \seel MURK\. From Tohono O'odham (Papago) *si:l mek* 'burnt saddle', containing *si:l* (from Spanish *silla*) 'saddle' and *mek* 'burnt'.

SIL NAKYA (AZ, Pima Co.) \seel NAHK yuh\. From Tohono O'odham (Papago) *si:l naggia* 'hanging saddle', containing *si:l* (from Spanish *silla*) 'saddle' and *naggia* 'hanging'.

SINYALA Canyon (AZ, Coconino Co.) Named for a Havasupai leader of the nineteenth century; also written **Sinyella**.

SIOVI SHUATAK (AZ, Pima Co.) From Tohono O'odham (Papago) *si'i'owĭ su:dagĭ*, lit. 'sweet water'.

SIPAPU \see pah POO\. The English term refers to an opening in the kiva, or ceremonial chamber, of the Pueblo peoples, which is believed to connect with the supernatural world. It may be derived from Keresan *šíp'áap'i*, but related words occur in other Pueblo languages. As a placename, Sipapu is found in Arizona (Coconino Co.) and New Mexico (Taos Co.).

SIPAULOVI (AZ, Navajo Co.) \si POL uh vee\. A variant writing of **SHIPAULOVI**.

SISNATHYEL Mesa (NM, Sandoval Co.) From Navajo *sis naateel* 'wide belt', containing *sis* 'belt', *naa-* 'across', and *-teel* 'wide'.

SITTING BULL. This was the English name of the Lakota chief *Tatanka Yotanka*, lit. 'buffalo-bull sitting', containing *thathą́ka* 'buffalo bull' and *íyotaka* 'to sit down'. As a placename, **Sitting Bull** is found in several states, such as New Mexico (Eddy Co.).

SITTING GOD (AZ, Apache Co.) From Navajo *yé'íí dah sidáí* 'the god sits at an elevation', containing *yé'íí* 'divine being'.

SITTING ROCK (AZ, Canyon de Chelly, Apache Co.) A translation of Navajo *tsé sí'ání* 'rock sits', containing *tsé* 'rock'.

SIWILI CHUCHG (AZ, Pima Co.) \SEE wi li CHOOCH'g\. From Tohono O'odham (Papago) *siwol cu:ck* 'onions standing', containing *siwol* 'onion' (from Spanish *cebolla*) and *cu:ck* '(plural) to be standing'.

SIWUKVA (AZ, Navajo Co.) Represents the Hopi placename *siwukva,* perhaps from *siwuk* 'solids fall out' and *va, paahu* 'water, spring'.

SKEDEE (OK, Pawnee Co.) \SKEE dee, SKID ee, ski DEE\. Named for the *Skidi* band of the Pawnee people.

SKIATOOK (OK, Osage Co.) \SKIE uh took\. Named for *Skiatooka,* an Osage leader. Near the town are **Skiatook Lake** and the **Skiatook Wildlife Management Area.**

SKOKAONAK (AZ, Pima Co.) \skoh KAH nuk\. From Tohono O'odham (Papago) *skokawagk* 'many staghorn cactus', with *kokaw* 'staghorn cactus'.

SKOKSONAK (AZ, Pima Co.) \skohk SOH nuk\. From Tohono O'odham (Papago) *skoksonag* 'full of woodrats', containing *ksona* 'woodrat'.

SKULLYVILLE (OK, Le Flore Co.) \SKUL ee vil\. From Choctaw, said to mean 'money town'; related to *iskali, skali* 'a bit (12 1/2¢), a dime; money', perhaps from French *escalin* 'shilling'. From 1864 to 1907, **Skullyville County** was part of **MOSHULTUBBEE District** in the Choctaw Nation.

SKUNK. The name of this native American mammal is from Massachusett Algonquian, derived from ancient Algonquian **šeka:kwa,* from **šek-* 'to urinate' and **-a:kw* 'fox'. There is a **Skunk Creek** in Oklahoma (Tulsa Co.) and another in Texas (Lipscomb Co.).

SLEEPING DUCK Rock (AZ, Apache Co.) A translation of Navajo *tsé naal'eełí sitíní* 'lying-down duck rock', containing *tsé* 'rock' and *naal'eełí* 'duck, goose'.

SLIDING OFF Rock (AZ, Apache Co.) Corresponds to Navajo *tsé naashzhoojí* 'sliding-off rock', containing *tsé* 'rock'.

SLIDING ROCK Ruins (AZ, Apache Co.) A translation of Navajo *kin náázhoozhí* 'the house that slid down', containing *kin* 'house, building' and *náázhoozh* 'it slid down'.

SOFKA (OK, Creek Co.) \SOF kuh\. In the traditional Creek Nation. Perhaps shortened from Muskogee *hvcce-sofke,* the name of a tribal town; this apparently consists of *hvcce* 'river' and *osafke, safke* 'sofkee, sofkey, a type of corn gruel'.

SOHU Park (AZ, Maricopa Co.) \SOH hoo\. From Hopi *soohu* 'star'.

SOLDANI (OK, Osage Co.) Named for Sylvester Soldani, a prominent Osage.

SONOITA (AZ, Santa Cruz Co.) \suh NOI tuh\. From Tohono O'odham (Papago) *şon oidag* 'spring field', from *şon* 'a spring' and *oidag* 'field'. The related form **Sonoyta** also occurs (AZ, Pima Co.).

SONOMA (TX, Ellis Co.) A transfer from *Sonoma County,* California; apparently from the name of an Indian group called *sonomas* or *sonomi,* of unclear origin.

SONORA (TX, Sutton Co.) Named for the state in northwestern Mexico; the term originally referred to a subgroup of the Ópata Indian tribe.

SONOYTA (AZ, Pima Co.) \suh NOI tuh\. A placename related to **SONOITA**.

SONSELA Buttes (AZ, Apache Co.) \sun SEL uh\. From Navajo *sǫ' silá* 'a pair of sitting stars', from *sǫ'* 'star' and *silá* 'they two sit'. A variant spelling is **Sunsela**.

SOPORI Wash (AZ, Santa Cruz Co.) Perhaps from *Sobaipuri,* the name of a Tohono O'odham (Papago) subgroup.

SOWATS Canyon (AZ, Coconino Co.) Said to be from Southern Paiute *showap* 'tobacco', but the correct Paiute form is *kwo'api.*

SPANISH Trail (AZ, Apache Co.) Corresponds to Navajo *naakaii adáánání ha'atiin,* lit. 'where Spaniards descended trail-up-out', containing *naakaii* 'Spaniard, Mexican'.

SPIDER Rock (AZ, Apache Co.) Corresponds to Navajo *tsé na'ashjé'ii* 'spider rock', containing *tsé* 'rock' and *na'ashjé'ii* 'spider'.

SPOKOGEE (OK, Hughes Co.) \spoh KOH gee\. In the historic Creek Nation; the place is a former townsite, now known as Dustin, near Wetumka. Its name was derived from the Muskogee placename *espokoke.*

SQUARE Butte (AZ, Coconino Co.) Corresponds to Navajo *tsé dik'ání* 'square rock', containing *tsé* 'rock'.

SQUAW Mountain (TX, Jack Co.) This word, referring to an Indian woman, was borrowed from New England Algonquian *squa* 'woman'. Over the years it has come to have a derogatory sense, and many Native Americans now consider it offensive because it means 'vulva' in Mohawk. In Arizona, **Squaw Peak,** near Phoenix, was renamed **PIESTEWA Peak**, after Pfc. Lori Piestewa, a Hopi woman who died in the Iraq War—the first Native American servicewoman ever to die in action. The placename **Squaw Tit** also occurs in Arizona (Maricopa Co.). **Squaw Creek,** in Lawton, Oklahoma, was recently renamed **NUMU Creek** (from Comanche *nʉmʉ* 'Comanche, people') at the request of Comanche tribe members.

STAN SHUATUK (AZ, Pima Co.) \stahn SHOO tuk\. From Tohono O'odham (Papago) *ston ṣu:dagĭ,* lit. 'hot water'.

STANDING COW Ruin (AZ, Apache Co.) A translation of Navajo *béegashii siziní kits'iil* 'standing cow ruin', containing *béegashii* 'cow', from Spanish *vacas* 'cows'.

STANDING RED ROCKS (AZ, Apache Co.) A translation of Navajo *tsé łichii 'íí'áhí* '(two) standing red rocks', containing *tsé* 'rock'.

STANDING ROCK (AZ, Apache Co.) A translation of Navajo *tsé 'íí'áhí* 'standing rock', containing *tsé* 'rock'. The name also occurs in New Mexico (McKinley Co.).

STINKING Lake (NM, Rio Arriba Co.) Probably a translation of the Tewa name, meaning 'smelly lake'.

STOA PITK (AZ, Pima Co.) \TOO wah peet, TOO wah peetk\. From Tohono O'odham (Papago) *stoha bidk* 'white mud', from *stoha* 'white' and *bidk* 'mud'.

STOA TONTK Well (Pima Co.) From O'odham *stoha ta:nk* 'white bank'.

STOA VAYA (Pima Co.). \TOO wah VWAH ya\. From O'odham *stoha wahia* 'white well'.

STOTONIC (AZ, Pinal Co.) From Tohono O'odham (Papago) *stotoñigk* 'many ants', from *totoñ* 'ant'. **Stotonyak** (AZ, Pima Co.) is from the same source.

STREAKED Rock (AZ, Apache Co.) From Navajo *tséndíłkéłí* 'streaked rock', containing *tsé* 'rock'.

SUAGEE (OK, Delaware Co.) In the historic Cherokee Nation; named for Wilson Suagee, a Cherokee leader.

SUGARITE (NM, Colfax Co.) \SHOOG uh reet\. This name has nothing to do with sugar; its folk etymology says that it comes from Spanish **Chicorica,** the feminine of *Chicorico* 'small-rich', from an Apachean word for 'turkey'. See also **CHICORICA Creek.**

SUNE Well (AZ, Yuma Co.) \SHOO nee\. Named for Chico Sune, a leader of the Sand Papago; perhaps from Tohono O'odham *şon* 'spring'.

SUNSELA Saddle (AZ, Apache Co.) A variant writing of **Sonsela.**

SUNSHINE Spring (AZ, Apache Co.) A translation of Navajo *shą́ą́'tóhí* 'sunshine water'; containing *shą́ą́'* 'sunshine'.

SUPAI (AZ, Coconino Co.) \SOO pie\. The name of this village on the **Havasupai Indian Reservation** is a shortening of the tribal name; the longer word means 'blue-water people'. Supai is inaccessible by car; visitors must enter the area by foot, horseback, or helicopter.

SUPI OIDAK (AZ, Pima Co.) \SUP ee OI dak\. From Tohono O'odham (Papago) *s-he:pi oidag* 'cold field', from *she:pi* 'cold' and *oidag* 'field'.

SUWANEE (NM, Valencia Co.) \SWA nee\. A transfer from the name of the *Suwannee River* in Georgia and Florida, itself derived from the name of a onetime Cherokee village, *Sawani*. The name has been transferred to other states, probably because of the popularity of Stephen C. Foster's song "Old Folks at Home" ("Way down upon the Swannee River").

SUWUK TONTK (AZ, Pima Co.) \suh wuk TAHNK\. From Tohono O'odham (Papago) *swegĭ ta:nk* 'red bank', containing *swegĭ* 'red' and *ta:nk* 'bank'.

SUWUKI CHUAPO (AZ, Maricopa Co.) \suh wuk CHOO poh\. Represents Tohono O'odham (Papago) *swegĭ ce:po* 'red bedrock-mortar', containing *swegĭ* 'red' and *ce:po* 'bedrock mortar').

T

TAAWAKI (AZ, Navajo Co.) \TAH wah kee\. From Hopi *taawaki* 'house of the sun', from *taawa* 'sun' and *ki, kiihu* 'house'.

TABIRA Ruin (NM, Torrance Co.) \tuh BEER uh\. From Spanish *Tabirá,* the site of a Tompiro pueblo destroyed in Spanish colonial times. The Tompiro language is no longer spoken.

TACUBAYA (TX, Brooks Co.) \tah koo BAH yuh\. Represents a transfer of the name of a neighborhood in Mexico City, from Aztec *atlacuihuayan* 'place where one draws water', from *atlacui* 'to draw water from a well'.

TAH CHEE (AZ, Apache Co.) \TAH chee, tah SHAY\. From Navajo *táchii'* 'red streak running into water'.

TAH HA BAH Well (NM, McKinley Co.) \TAH hah bah\. Perhaps Navajo *táhabąąh* 'spacious shore, edge', from *tá-* 'water', *ha-* 'space, area', and *bąąh* 'shore'.

TAHLEQUAH (OK, Cherokee Co.) \TAL uh kwah\. In the historic Cherokee Nation; from *Talikwa* or *Tellico,* the name of a Cherokee town in Tennessee. **Historic Tahlequah** incorporates buildings dating as early as 1839, when this Oklahoma town became the capital of the Cherokee Nation. (Another story has the name invented in Oklahoma and meaning 'two is enough'.).

TAHOE Creek (OK, Caddo Co.) Probably a transfer from the name of Lake Tahoe in California and Nevada, from Washoe *dá'aw.*

TAHOKA (TX, Lynn Co.) \tuh HOH kuh\. Said to be from an unidentified Indian language; translated as 'clear water' or 'deep water'.

TAHUACHAL, Banco (TX, Cameron Co.) \tah wah CHAHL\. The name of this levee (*banco* in Spanish) is probably a Mexican Spanish word meaning 'possum-place', from *tahuache, tacuache, tlacuache* 'possum', from Aztec *tlacuatzin* 'little possum', from *tlacuatl* 'possum'. The nearby **Banco Tahuachalito** \tah wah cha LEE toh\ is the diminutive, meaning 'little possum-place'.

TAHUTA Point (AZ, Coconino Co.) \tuh HOO tuh\. Named for Tahota Jones, a Havasupai basketmaker; her Native name was *tahóta,* perhaps meaning 'something concealed'.

TAIBAN (NM, DeBaca Co.) \TIE ban\. Said to be a loan word from an unidentified Indian language.

TAJIQUE (NM, Torrance Co.) \tah HEE kay, tuh HEE kee\. A Spanish spelling, also written **Taxique,** for a Tiwa village; the Native name was *Tüsh-yit-yay.*

TALAHOGAN Canyon (AZ, Navajo Co.) \tal uh HOH gun\. From Navajo *táala hooghan* 'flat-topped hogan (native house)'.

TALALA (OK, Rogers Co.) \tuh LAH luh\. In the traditional Cherokee Nation; named for a Cherokee officer in the Civil War, known in English as Captain Talala; or perhaps from Delaware *talala* or *talalakw* 'white cedar tree'.

TALIHINA (OK, Le Flore Co.) \tal uh HEE nuh\. In the historic Choctaw Nation; from Choctaw *tali-hina* 'railroad', containing *tali* 'rock, metal' and *hina* 'road'. **Lake Talihina** is nearby in Latimer County.

TALIMENA State Park (OK, Le Flore Co.) In the historic Choctaw Nation; the name is a blend of **TALIHINA** and *Mena,* the name of a town in Arkansas. The Talimena Scenic Drive runs between the state park and Mena.

TALL Mountain (AZ, Navajo Co.) Corresponds to Navajo *dził* 'mountain' and *-neez* 'long, tall'.

TALOAH (OK, Atoka Co.) Perhaps from the same origin as **Talowah,** Mississippi, from Choctaw, perhaps meaning 'many rocks', from *tali* 'rock' and *lawal* 'many'; or perhaps a personal name, *taloowa* 'singer'.

TALOGA (OK, Dewey Co.) \tuh LOH guh\. From a Creek word meaning 'rock in the water', which originally referred to a rock marking the boundary of the Creek Nation.

TALOKA (OK, Haskell Co.) \tuh LO kuh\. This name is of unclear derivation, although it may be related to **TALOGA** and may contain *tali* 'rock'. **Toloka Creek** and **Taloka Township** are both located in the Choctaw Nation. **Little Taloka Creek** is nearby.

TALPA (NM, Taos Co.) \TAHL puh, TAL puh\. From Mexican Spanish *Talpa,* a placename in Jalisco, Mexico; this is a dialect variant of Aztec *tlalpan* 'on the earth, on the shore', from *tlalli* 'earth'. The name **Talpa** also occurs in Texas (Coleman Co.).

TALPACATE Creek (TX, Bee Co.) \TAL puh kat\. Perhaps from Karankawa, meaning 'tadpole'.

TAMAHA (OK, Haskell Co.) \tah MAH hah, TAH muh hah\. In the historic Choctaw Nation; perhaps from Choctaw *tamaha* 'town'. Possibly related is **Tamahaw Peak** (OK, Haskell Co.).

TAMPICO (NM, McKinley Co.) \tam PEE koh\. Named for a city in Tamaulipas state, eastern Mexico. The origin of the name is Huastec (Mayan) *tampik'o* 'place of dogs', from *pik'o* 'dog'. There is also a **Tampico** in Texas (Hall Co.).

TANDY (OK, Hughes Co.) \TAN dee\. In the historic Creek Nation; said to be named for Tandy Walker, a Cherokee leader.

TANO Point (NM, Santa Fe Co.) \TAH noh\. This Spanish form, from a probable Native form, *tõáno* or *tõá:nu,* was applied in colonial times to the Tewa Indians of the Rio Grande Valley. It is the origin of the name *Tanoan* \tuh NOH un\, which anthropologists apply to the language family that includes Tewa. A related placename is the Spanish plural **Los Tanos** \lohs TAH nos\ (NM, Guadalupe Co.).

TANTABOGUE Creek (TX, Trinity Co.) Perhaps from Choctaw *book* 'creek', but of unclear derivation.

TAOS (NM, Taos Co.) \TAH ohs\. The county seat of **Taos County**, which in turn was named for nearby **Taos Pueblo**, an ancient Native American village. The name of the pueblo and of the Tanoan Indian community is from the Native name *tôotho* 'in the village'; it is also said to mean 'red willow' in the Tiwa language. The local placename **Cerro de los Taoses** means 'hill of the Taos Indians'.

TAPEATS Creek (AZ, Coconino Co.) \TAP uts\. Said to have been the name of a Southern Paiute resident.

TAQUACHE Creek (TX, Zapata Co.) \tuh KWAH chee\. From Mexican Spanish *tlacuache* 'opossum', from Aztec *tlacuatzin.*

TARANCAHUAS Creek (TX, Duval Co.) \tuh RAHNG kuh wahs\. Perhaps a Spanish plural reflecting an alternative form of *Karankawa,* the name of an Indian group that once lived in Texas.

TASCALA Canyon (AZ, Santa Cruz Co.) \tah SKAH luh\. Perhaps an alternative form of Mexican Spanish *Tlaxcala,* the name of a city and state in central Mexico; from Aztec *tlaxcallan* 'place of tortillas', from *tlaxcalli* 'tortilla'.

TAT MOMOLI (AZ, Pinal Co.) \taht MOH moh li\. Pima *taḍ memelǐ* 'footrace', from *taḍ* 'foot' and *memelǐ* 'running'. A related placename in the same area is **Tat Momolikot Dam,** with *memelǐkudǐ* 'place of the footrace'.

TATAHATSO Wash (AZ, Coconino Co.) Perhaps from Navajo *bidáá hatsoh* 'big rim', with the English name influenced by the name of nearby **TATAHOYSA Wash,** which is also perhaps from Navajo but is of unclear derivation.

TATAHOYSA Wash (AZ, Coconino Co.) Perhaps from Navajo, but the derivation is unclear. See **TATAHATSO.**

TATAI TOAK (AZ, Pima Co.) From Tohono O'odham (Papago) *taḍai do'ag,* lit. 'roadrunner mountain'.

TATK KAM VO (AZ, Pima Co.) Tohono O'odham (Papago) *ta:tad wo'o* 'pond of the feet', from *ta:tad* 'feet' (*tad* 'foot') and *wo'o* 'pond'. An alternative form of the name is **Tatkum Vo.**

TATKUM VO (AZ, Pima Co.) An alternative form of **TATK KAM VO.**

TAUPA (OK, Comanche Co.) Said to be a Comanche personal name.

TAXIQUE. See **TAJIQUE.**

TAWA Point (AZ, Apache Co.) \TAH wuh\. From Hopi *taawa* 'sun'. There is also a **Tawa** in New Mexico (Rio Arriba Co.).

TAWAKONI, Lake (TX, Hunt and Rains Cos.) The name of a Caddoan subtribe, related to the Wacos and Wichitas. The Tawakoni people are among the **Wichita and Affiliated Tribes,** which have their headquarters at Anadarko, Oklahoma (Caddo Co.). A related placename is **TEHUACANA** (TX, Limestone Co.).

TAWAPA Spring (AZ, Navajo Co.) \tuh WAH puh\. From Hopi *taawapa* 'sun spring', containing *taawa* 'sun' and *pa, paahu* 'water, spring'. There is also a **Tawapa** in New Mexico (Sandoval Co.).

TECOLOTE (NM, Lincoln Co.) \tek uh LOH tee\. This Mexican Spanish word for 'owl' is from Aztec *tecolotl.* Related placenames are **TECOLOTEÑOS** and **TECOLOTITO.**

TECOLOTEÑOS (NM, San Miguel Co.) \tek uh loh TANE yohs\. From *tecolote,* the Mexican Spanish word for 'owl', from Aztec *tecolotl.* This form of the word means 'little owl-people'.

TECOLOTITO (NM, San Miguel Co.) \tek uh loh TEE toh\. Appears to be the Spanish diminutive of **TECOLOTE**—that is, 'little owl'. However, it more likely means 'the small place founded as an offshoot of the place called Tecolote'.

TECUMSEH (OK, Pottawatomie Co.) \tuh KUM suh, tuh KUM see\. Named for the renowned Shawnee Indian war chief who fought with the British in the War of 1812 and was commissioned as a brigadier general in the British army. His Native name may have been *tkamʔθe* 'going across'; the Menominee equivalent is *tahkamehsɛːw* 'flies straight across'. Near the Oklahoma town is **Tecumseh Lake.**

TEE PEE \TEE pee\. The term is used for a type of house, made of poles and animal hides, characteristic of Plains Indian culture; the origin is Dakota *thípi* 'house'. There are a **Tee Pee Creek** and a **Tee Pee City** in Texas, both in Motley County. Alternative spellings are *Teepee, Tepee,* and *Tipi.*

TEEC-NI-DI-TSO Wash (NM, San Juan Co.) \teek ni DIT soh\. From Navajo *t'iis niditso* 'cottonwood trees beginning to descend yellow', containing *t'iis* 'cottonwood', *ni-* 'earth', *di-* 'beginning', and *-tso* 'yellow, light green'.

TEEC NOS POS (AZ, Apache Co.) \teek nohs POS, tees nohs POS\. From Navajo *t'iis názbąs* 'cottonwoods in a circle', from *t'iis* 'cottonwood trees' and *názbąs* 'circle, circular'.

TEESQUATNEE (OK, Delaware Co.) \TEES kwat nee\. In the Cherokee Nation; perhaps the Cherokee name of a person who lived there.

TEES TOH (AZ, Navajo Co.) \tees TOH\. From Navajo *t'iis tó,* lit. 'cottonwood water'.

TE'EWI (NM, Rio Arriba Co.) Also written **Teewi;** an abandoned pueblo once occupied by speakers of the Tewa language.

TEGUA (AZ, Navajo Co.) \TAY gwuh, TAY wuh\. From Spanish *Tegua,* referring to the Tewa Indians who live on the Hopi First Mesa. An alternative spelling is **Tequa.** The names **Tegua, TEHUA Hills,** and **TEWA** all have the same origin.

TEHUA Hills (AZ, Navajo Co.) \TAY wuh\. From Spanish *Tegua,* referring to the *Tewa* Indians who live on the Hopi First Mesa. The Spanish spelling comes from the same origin as **TEGUA** and **TEWA.**

TEHUACANA (TX, Limestone Co.) \ti WA kuh nuh\. A Spanish spelling of the name of a Wichita subgroup, whose name is also written *Tawakoni* and *Tewockony.* **Tehuacana Creek** flows into the Frio River (TX, Frio Co.). A related placename is **Lake TAWAKONI** (TX, Hunt and Rains Cos.).

TEKAPO Reservoir (NM, McKinley Co.) On the Zuni Reservation; perhaps from Zuni *tek'appowa* 'hill'.

TENABO (NM, Torrance Co.) \TEN uh boh\. From Spanish *Tenabó,* from the name of an abandoned Tompiro pueblo.

TENAHA (TX, Shelby Co.) \TEN uh haw\. Said to mean 'muddy water' in an unidentified Indian language, but another possible source is Spanish *tinaja* 'large jar; rock waterhole'. **Tenaha Creek** flows into the Sabine River (Shelby Co.).

TENKILLER Ferry Lake (OK, Sequoyah Co.) In the Cherokee Nation; named for a Cherokee family that lived nearby. The term is a war name referring to an ancestor who killed ten enemies in battle. Often simply called **Tenkiller Lake** by locals. **Tenkiller Wildlife Management Area** and **Tenkiller State Park** are nearby.

TENNESSEE Colony (TX, Anderson Co.) \TEN uh see\. Named for the home state of some of its original settlers, which was named (as was the river) for a Cherokee village, *tănăsí, tănasí,* of no known etymology.

TEQUA (AZ, Navajo Co.) Probably a misspelling of Spanish **TEGUA** or **TEHUA,** referring to the **TEWA** Indians.

TEQUESQUITE Creek (NM, Harding Co.) \tek us KEE tee\. Mexican Spanish for 'alkali, saltpeter', from Aztec *tequixquitl.* The related name **Tequisquite Creek** also occurs in Texas (Maverick Co.). **TESESQUITE Creek** in New Mexico (Union Co.) and Oklahoma (Cimarron Co.) may be related.

TERLINGUA (TX, Brewster Co.) \ter LING gwuh, ter LING guh\. Perhaps from an Indian language, but perhaps from Spanish *tres lenguas* 'three languages', referring to the Spanish, English, and Indian languages.

TERRAPIN Creek (TX, Nacogdoches Co.) This word for a kind of turtle was borrowed from Virginia Algonquian; consider the related Munsee *tó:lpe:w.*

TES NEZ IAH (AZ, Apache Co.) \tees nez EE yah\. From Navajo *t'iis nééz íí'á* 'tall cottonwood stands', containing *t'iis* 'cottonwood tree'.

TES YAHZ LANI (AZ, Coconino Co.) \tees yahz LAH nee\. From Navajo *t'iis yáázh łání* 'many little cottonwoods', containing *t'iis* 'cottonwood tree'.

TESBITO (AZ, Navajo Co.) \TEES bi toh, tes BEE toh\. From Navajo *t'iis bitó* 'cottonwood spring' (lit. 'cottonwood its-water'), containing *t'iis* 'cottonwood tree', *bi-* 'its', and *tó* 'water, spring'.

TESESQUITE Creek (NM, Union Co.; OK, Cimarron Co.) This stream flows westward into the Cimarron River. The name may be an error for **Tequesquite**, Mexican Spanish for 'alkali'.

TESHIM Butte (AZ, Navajo Co.) From Navajo, of unclear derivation.

TESUQUE (NM, Santa Fe Co.) \ti SOO kee\. From Tewa *t'athuŋye,* perhaps 'down at the dry spotted place', from *t'a* 'dry', *thu* 'spotted', and *-ye* 'down at'.

TEWA \TAY wuh\. The name refers to speakers of a Tanoan language who originally lived in several pueblos, such as Santa Clara and Ohkay Owingeh (San Juan), in the Rio Grande Valley of New Mexico. After the Pueblo Revolt of 1680, some Tewas fled westward and settled among the Hopis in Arizona; this group is now called "Hopi-Tewa." As a placename, **Tewa** refers to a settlement in Arizona (Navajo Co.); related names are **TEHUA Hills** and **TEGUA**.

TEXANA (TX, Jackson Co.) \tek SAN uh\. This first town in the county was given the name Santa Anna when founded in 1832, referring to Antonio López de Santa Anna, a nineteenth-century Mexican leader; it was later changed to Texana in honor of the independence of Texas. The town no longer exists, but the name is borne by **Lake Texana.** A related placename is **Texanna** (OK, McIntosh Co.).

TEXARKANA (TX, Bowie Co.) \teks ahr KAN uh\. The name was created by combining elements from the placenames *Texas, Arkansas,* and *Louisiana.*

TEXAS (state) \TEKS uhs\. From Spanish *Tejas,* used by the Spaniards to refer to Indian peoples also called *Teyas;* from Caddo *táyša'* 'friend, ally'. The name **Texas** occurs in transfer names in other states—for example, **Texas Mine** (NM, Rio Arriba Co.), **Texas Hill** (AZ, Yuma Co.), and **Texas County** (OK).

TEXHOMA (OK, Texas Co.; TX, Sherman Co.) \teks HOH muh\. A blend of the placenames *Texas* and *Oklahoma.* **Lake Texoma** is on the border of Texas and Oklahoma.

TEXICO (NM, Curry Co.) \TEKS i koh\. A blend of the placenames *Texas* and *New Mexico.*

TEXLA (TX, Orange Co.) \TEKS luh\. A blend of elements from the placenames *Texas* and *Louisiana;* the town is near the state line.

TEXLINE (TX, Dallam Co.) A blend of *Texas* and *line;* so named from being close to the New Mexico state line.

TEXMO (OK, Roger Mills Co.) \TEKS moh\. A blend of elements of *Texas* and *Missouri*.

TEXOKLA (OK, Beckham Co.) \teks OH kluh\. A blend of *Texas* and *Oklahoma*. Similar nearby names are **Texola** \teks OH luh\ and **Lake Texoma** \teks OH muh\, which lies on the border between the two states.

TEXOWA (OK, Tillman Co.) A blend of the placename *Texas* and the name of the **KIOWA** Indian people, whose tribal headquarters is at Carnegie, Oklahoma.

TEYPAMA (NM, Socorro Co.) \ti PAH muh\. An archaeological site once occupied by the Piro Indians; also written **Teypana**.

TEZAH Mountain (NM, McKinley Co.) \TEZ ah\. Probably from Navajo *deez'á* 'it extends as a ridge, bluff, point'.

THINGS FLOW AROUND THE STONE (AZ, Apache Co.) Corresponds to Navajo *tsé biná'áz'éí* 'rock around which floating took place'.

THLOPTHLOCCO (OK, Okfuskee Co.) \thlop THLUK oh\. In the historic Creek Nation; from the Muskogee placename *raprakko*. This Muskogee community, known in full as the **Thlopthlocco Tribal Town of the Creek Nation of Oklahoma,** is located at the town of **WELEETKA**.

THOHEDLIH (NM, San Juan Co.) From Navajo *tó aheedlį* 'waters converge, flow together', from *tó* 'water', *ahee (ahi)* 'convergent', and *(d)lį* 'flow'.

THUNDERBIRD (NM, Torrance Co.) Among many North American Indian peoples, thunder was said to be caused by this legendary bird; the English word corresponds to Native terms such as Ojibwa *animikii*. There is a **Thunderbird Park** in Arizona (Maricopa Co.) and places named **Lake Thunderbird** in Oklahoma (Cleveland Co.) and Texas (Bastrop Co.).

TIA JUANA River (OK, Delaware Co.) \tee uh WAH nuh\. Transferred from the city of *Tijuana* in Baja California, Mexico. From the name of a Diegueño village, written as *Tiajuan* in 1829. The name of the city has often been Anglicized as *Tía Juana* 'Aunt Jane'.

TIAK Wildlife Management Area (OK, McCurtain Co.) In the historic Choctaw Nation; from Choctaw *tiyak* 'pine tree'.

TIAWAH (OK, Rogers Co.) \TIE uh wah, TIE wah\. In the historic Cherokee Nation; said to be named for an Indian mound in Georgia, but the language of origin is uncertain.

TIERRA AMARILLA (NM, Rio Arriba Co.) \ti YARE uh ah mah REE yuh\. The Spanish name, meaning 'yellow earth', corresponds to Navajo *łitsooí* 'yellow place,' and the Tewa name for the place has the same meaning.

TIGER (OK, Creek Co.) In the historic Creek Nation; named for Billy Tiger, a prominent Muskogee. The English word *tiger* is used locally to mean 'cougar, mountain lion', corresponding to Muskogee *kaccv*.

TIGUA Pueblo (TX, El Paso Co.) \TEE wuh\. An alternative name for **YSLETA (del Sur)**, a settlement of Tiwa Indians who originally lived in **ISLETA** (NM, Bernalillo Co.) The **Tigua Reservation** is at El Paso, Texas. A related but distinct Indian group is the **TEWA**.

TIMP Canyon (AZ, Coconino Co.) \timp\. From Southern Paiute *tïmpi* 'rock'.

TIOGA (TX, Grayson Co.) \tie OH guh\. Transferred from Tioga County in New York state. From Mohawk *teyó:kę* 'junction or fork' or *teyaó:kę* 'having a junction or fork', from the root *-okę-* 'to be Y-shaped'.

TISHOMINGO (OK, Johnston Co.) \tish uh MING goh\. Named for a Chickasaw leader whose Native name was *tisho' mįko*, containing *tisho'* 'assistant to an Indian doctor' and *mįko* 'chief'. The town was the seat of **Tishomingo District** in the Chickasaw Nation between 1866 and 1907 and is the historic capital of the Chickasaw Nation. Nearby is the **Tishomingo National Wildlife Reserve.**

TITHUMIJI (AZ, Coconino Co.) The term is the surname of a local Havasupai family.

TIYO Point (AZ, Coconino Co.) \TEE yoh\. From Hopi *tiyo* 'young man'.

TIZ NA TZIN Trading Post (NM, San Juan Co.) \tiz NAH tsin\. Perhaps from Navajo *t'iis náshjin* 'black cottonwood circle'.

TOADLENA (NM, San Juan Co.) \toh ad LEE nuh, tohd LEE nuh\. From Navajo *tó háálį́* 'it flows up and out', from *tó* 'water, spring', *há(á-)* 'up and out', and *lį́* 'it flows'.

TOBACCO Creek (TX, Borden Co.) The name of the herb widely used for smoking is from Spanish *tabaco*, borrowed from an Arawakan language of the Caribbean.

TO BILA'I (NM, San Juan Co.) From Navajo *tó bíla'í* 'water fingers'.

TO-BIL-HASK-IDI Wash (NM, San Juan Co.) From Navajo *tó bił hask'idí* 'water with a hill', containing *tó* 'water', *bił* 'with it', and *hask'id* 'it mounds, makes a hill'.

TOBOXKY (OK, Pittsburg Co.) \tuh BOKS kee\. In the historic Choctaw Nation; from Choctaw *tobaksi* 'coal'. A variant spelling is found in the name **Tobucksy County**, which was the western section of **MOSHULATUBBEE District** in the Choctaw Nation between 1864 and 1907.

TOCITO (NM, San Juan Co.) \toh SEE toh\. A Spanish-style spelling for Navajo *tó sido* 'hot spring, hot water', from *tó* 'water' and *sido* 'it is hot'.

TODACHEENE Lake (NM, San Juan Co.) \toh duh CHEE nee\. On the Navajo Nation; from Navajo *tó dích'íi'nii* 'bitter water, alkaline water', containing *tó* 'water' and *dích'íí'* 'it is bitter, sour'.

TODASTONI Spring (AZ, Apache Co.) A variant spelling of **TOH DAHSTINI Spring.**

TO-DIL-HIL Wash (NM, San Juan Co.) \toh DEEL heel\. From Navajo *tódiłhił* 'dark water, whiskey', from *tó* 'water' and *diłhił* 'it is dark'. Also called "Whiskey Creek" in English. There is a **WHISKEY Creek** in Arizona (Apache Co.).

TODILTO Park (NM, McKinley Co.) \toh DEEL toh\. From Navajo *tó dildǫ'* 'water that pops, sounding water', from *tó* 'water, spring' and *dildǫ'* 'it pops, explodes'. A related name is **Tohdildonih Wash** (NM, McKinley Co.).

TODOKOZH (AZ, Navajo Co.) A variant spelling of **TOH DE COZ**.

TOE Rock (NM, Sandoval Co.) Said to be a translation of a Tiwa placename.

TOGOYE Lake (NM, McKinley Co.) Probably from Navajo *tóyéé'* 'scarce water', from *tó* 'water' and *yéé'* 'scarce'.

TOH AH CHI (AZ, Navajo Co.) \toh HAH chee\. From Navajo *tóhaach'i'* 'water is scratched out', from *tó* 'water' and *haach'i'* 'it is scratched out'. Related placenames are **TOHATCHI** and **TOHACHE**.

TOH ATIN Mesa (AZ, Apache Co.) \toh AH tin\. Shortened from Navajo *tó ádin dah azká* 'no-water mesa', from *tó* 'water', *ádin* 'non-existent', *dah* 'above', and *asgą'* 'something dried up'.

TOH CHIN LINI Canyon (AZ, Apache Co.) \toh chin LEE nee\. From Navajo *tó ch'ínlíní* 'water flowing out'.

TOH DAHSTINI Spring (AZ, Apache Co.) \toh dah STEE nee\. Perhaps a contraction from Navajo *tó dah da'aztání* 'fingers of water here and there above'. Variant spellings are **Todastoni** and **Tohdenstani Spring**.

TOH DE COZ (AZ, Navajo Co.) \TOH di kohz\. \toh di KOOZ\. From Navajo *tó dík'ǫ́ǫ́zhí* 'bitter water', containing *tó* 'water' and *dík'ǫ́ǫ́zh* 'salty, sour, bitter'. A variant spelling is **Todokozh**.

TOHACHE Wash (NM, San Juan Co.) \toh HACH ee\. From Navajo *tóhaach'i'* 'water is scratched out', from *tó* 'water', and *haach'i'* 'it is scratched out'. The name **Tohache** also occurs in Arizona (Apache Co.), where it is pronounced \toh HAH chee\. Related placenames are **TOH AH CHI** and **TOHATCHI**.

TOHASGED Spring (NM, McKinley Co.) Navajo *tó haasgeed* 'water is dug out', from *tó* 'water' and *haasgeed* 'is dug out'.

TOHATCHI (NM, McKinley Co.) \TOH hach ee\. From Navajo *tóhaach'i'* 'water is scratched out', from *tó* 'water' and *haach'i'* 'it is scratched out'. Related names are **TOH AH CHI** and **TOHACHE Wash**.

TOHDENSTANI Spring (AZ, Apache Co.) A variant spelling of **TOH DAHSTINI Spring**.

TOHDILDONIH Wash (NM, McKinley Co.) From the same origin as **TODILTO**.

TOHEE Township (OK, Lincoln Co.) \TOH hee\. Named for a leader of the Iowa tribe. It is unclear whether the name has anything to do with *towhee,* the name of a bird.

TOHLAKAI (NM, McKinley Co.) \toh luh KIE\. From Navajo *tó łigaaí* 'white water', from *tó* 'water, spring' and *łigaaí* 'it is white'.

TOHNALI Mesa (AZ, Coconino Co.) \toh nah LEE\. From Navajo *tó náálíní* 'water flowing downward'.

TOH-NI-TSA Lookout (NM, San Juan Co.) \TOH nit sah\. From Navajo *tónitsaa* 'big water', from *tó* 'water' and *nitsaa* 'big'.

TOH-NOZ-BOSA Well (NM, McKinley Co.) \toh NOHZ buh suh\. From Navajo *tó názbąsí* 'round water, circular water', containing *tó* 'water' and *názbąs* 'round, circular'.

TOHONO O'ODHAM Indian Reservation (AZ, Pima Co.) \TOH hoh noh OH ohd hahm\. The name refers to the tribe formerly called **Papago**; it means 'desert people', by contrast with *Akimel O'odham* 'river people', referring to the closely related **PIMA** tribe. Both groups speak languages in the Piman group of the Uto-Aztecan language family. The Tohono O'odham also live on the **SAN XAVIER DEL BAC** and the **GILA** Bend Indian Reservations. *O'odham* is sometimes used as a cover term for both tribes or both languages together.

TOKIO (OK, Kiowa Co.) Said to be from Kiowa *jócyói* 'long house'. The same name in Terry Co., Texas, may be from an alternative spelling of *Tokyo,* the city in Japan.

TOL DOHN Spring (NM, McKinley Co.) \tohl DOHN\. Perhaps from the same origin as **TODILTO Park** (NM, McKinley Co.).

TOLANI Lake (AZ, Coconino Co.) \toh LAH nee\. From Navajo *tółání* 'much water, many bodies of water', from *tó* 'water" and *łání* 'many, much'.

TOLAR (NM, Roosevelt Co.) Perhaps from Mexican Spanish *tular* 'a patch of tules or cattail reeds', from *tule* 'cattail reed, tule', from Aztec *tolli.* Another name from the same source is **El Tular** (TX, Cameron Co.).

TOLCHACO Gap (AZ, Coconino Co.) Probably a variant spelling of **TOLCHICO**.

TOLCHICO (AZ, Coconino Co.) \tohl CHEE koh\. From Navajo *tółchí'íkooh* 'red-water wash', from *tó* 'water', *-łchí'í, łichíí* 'red', and *kooh* 'a wash, an arroyo'.

TOLOKA (OK, Haskell Co.) \tuh LOH kuh\. Perhaps from Choctaw, but the derivation is unclear.

TOLTEC (NM, Cibola Co.) \TOL tek\. From Aztec *toltecah,* designating an ancient people of Mexico. The literal meaning is 'people of the tules or cattail reeds'.

TOLUCA (TX, Hidalgo Co.) \toh LOO kuh, tuh LOO kuh\. A transfer name from a city in Mexico, orginally *Tolocan* in Aztec, meaning 'place of (the god) Tolo'. The Texas post office and town were later renamed Progreso.

TOMB OF THE WEAVER (AZ, Apache Co.) Corresponds to Navajo *hastiin tł'ó bitséyaa* 'mister weaver's cave'.

TONALEA (AZ, Coconino Co.) \toh nuh LEE uh\. From Navajo *tó nehelį́į́h* 'water flows to a point', from *tó* 'water', *nehe, nihi* 'to a point', and and *lį́į́h, lį́* 'flow'.

TO-NIL-CHONI Wash (NM, San Juan Co.) From Navajo *tónitchxoní* 'stinking water', containing *tó* 'water' and *nitchxon* 'it stinks'.

TONK Creek (TX, Throckmorton Co.) An abbreviation of *Tonkawa,* the name of a Texas Indian people.

TONKAWA (OK, Kay Co.) \TONG kuh wuh, TONG kuh wah, TONG kuh way\. Perhaps derived from Waco *tonkawéya* 'they all stay together'. The town is the headquarters of the **Tonkawa Tribe.** The name also occurs in Texas (Cooke Co.), and a related placename is **Tonkaway Lake** (TX, Brazos Co.).

TONOKA Valley (AZ, Pima Co.) \TOH nuh kuh\. From Tohono O'odham (Papago) *to:nk* 'dike, hill'.

TONOPAH (AZ, Maricopa Co.) \TOH nuh pah\. Perhaps a transfer from Nye County., Nevada. Derived from Shoshoni *tonoppaa* 'greasewood spring', from *tonoppi* 'greasewood' and *paa* 'water, spring'.

TONQUA Creek (TX, DeWitt Co.) Perhaps from the name of the **TONKAWA** Indian people.

TONQUE, Arroyo (NM, Sandoval Co.) \TONG kay\. Probably from Spanish *Tunque,* the name of a precolonial pueblo, perhaps in turn from Tewa *t'ųye* 'down at the basket', combining *t'ųŋy* 'basket' and *-ye* 'down at'. A variant spelling appears in the name **Tunque Pueblo.**

TONTO (AZ, Gila Co.) \TON toh\. This Spanish word means 'foolish', but reference here is probably to *Tonto Apache,* the term formerly used to refer to the Western Apaches. The older name probably represents a Spanish translation of Chiricahua Apache *bini:'édiné* or Mescalero Apache *bini:'édinendé* 'people without minds', terms they used to designate the Western Apaches. The **Tonto Apache Tribe** is a small Western Apache group living around Payson, Arizona (Gila Co.). **Tonto Bridge National Forest** and **Tonto Natural Bridge State Park** are in the same area.

TONUCO (NM, Doña Ana Co.) \tuh NOO koh\. Perhaps an Indian placename, language undentified.

TONUK VO (AZ, Pima Co.) From Tohono O'odham (Papago) *to:nk wo'o,* lit. 'dike pond'.

TOPACOBA Hilltop (AZ, Coconino Co.) A spelling variant of **TOPOCOBA.**

TOPAWA (AZ, Pima Co.) \toh PAH wuh, tuh PAH wuh\. Perhaps from Tohono O'odham (Papago) *oḍ bawe* 'it is a tepary bean'.

TOPOCOBA Hilltop (AZ, Coconino Co.) From the Havasupai placename *Tovokyóva*. Also spelled **Topacoba**.

TOREVA (AZ, Navajo Co.) \toh REE vuh, tuh REE vuh\. Corresponds to Hopi *toriva paahu* 'twisted spring', from *tori* 'twist' and *va, paahu* 'water, spring'.

TORONTO (TX, Brewster Co.) \tuh RON toh\. Allegedly from an Indian word meaning 'rising land'; the town was in a mountain pass near Alpine, Texas. Or perhaps this is a transfer name from Toronto in Ontario, Canada. The name of the Canadian city is from an Iroquoian language, but the derivation is unclear.

TOROWEAP Valley (AZ, Mohave Co.) \tore uh WEEP\. Probably from Southern Paiute *tutu-uippi* 'whirling canyon', from *tutu* 'to whirl' and *uippi* 'canyon'. A variant form is **Tuweep**.

TOTACON (AZ, Apache Co.) From Navajo, meaning 'sweet water', with *tó* 'water' and *łikan* 'be sweet'.

TOTASKWINU Ruin (NM, Sandoval Co.) From a Jemez (Towa) placename, *tootakwinun,* perhaps containing *-kwi* 'to be standing' and *-nun* 'place'.

TOTAVI (NM, Santa Fe Co.) Perhaps from a Tewa word meaning 'quail'.

TOTONTEAC Mountain (AZ, Maricopa Co.) \toh TON tee ahk\. Originally, the name *Totonteac* was assigned by Spanish explorers to what is now Arizona; it is perhaps from Aztec *totonqui* 'hot'.

TOTOPITK (AZ, Maricopa Co.) Perhaps from Tohono O'odham (Papago) *top't* 'crooked, lopsided'.

TOVAKWA Ruin (NM, Sandoval Co.) From the Jemez (Towa) placename *túuva-kwa*.

TOWAGO Point (AZ, Coconino Co.) The surname of a Havasupai family, perhaps from *tahwága* 'to have two wives'.

TOYAH (TX, Reeves Co.) Said to mean 'flowing water' in an unidentified Indian language, but perhaps from Comanche *toya* 'mountain'. Nearby **Toyah Creek** flows into Balmorhea Lake.

TOYEI (AZ, Apache Co.) \toy AY\. From Navajo *tó-yéé,* lit. 'water-scarce'. A related name is **Tuye Spring** (AZ, Apache Co.).

TRAIL THE MEXICANS CAME DOWN Canyon (AZ, Apache Co.) From Navajo *naakaii adáánání* 'trail the Mexicans came down', containing *naakaii* 'Spaniard, Mexican'.

TRAIL WHERE THE ENEMY WALKED UP SINGING (AZ, Apache Co.) From Navajo *anaa' sin yił haayáhí* 'where the enemy (a Hopi woman) walked up singing', containing *anaa'* 'enemy'.

TSA LA GI Indian Village (OK, Cherokee Co.) \tsah lah GEE, sah lah GEE\. In the historic Cherokee Nation; from Cherokee *jalagi* 'Cherokee'.

TSAH TAH Trading Post (NM, San Juan Co.) \TSAH tah\. From Navajo *ts'ahtah* 'among the sagebrush', from *ts'ah* 'sagebrush, *Artemesia*' and *-tah* 'among'.

TSAI SKIZZI Rock (AZ, Coconino Co.) \say SKEE zee\. From Navajo *tsé k'izí* 'rock cleft', containing *tsé* 'rock'.

TSAILE (AZ, Apache Co.) \say LEE\. From Navajo *tsééhílį́* 'it flows into a canyon', from *tsééh* 'into rock, into a canyon' and *(y)ílį́* 'it flows'.

TSAMA Archaeological District (NM, Rio Arriba Co.) A variant writing of **CHAMA,** from Tewa *tsąmą* 'to wrestle'.

TSANKAWI Ruins (NM, Santa Fe Co.) \TSANG kuh wee\. From Tewa *sǫk'wewi* 'gap of the sharp round-cactus', from *sǫ* 'round cactus', *k'e* 'sharp', and *wi'i* 'gap'.

TSAYA (NM, San Juan Co.) From Navajo *tséyaa* 'beneath the rock', from *tsé* 'rock' and *yaa* 'beneath'.

TSAYATOH (NM, McKinley Co.) From Navajo *tséyaa tó* 'rock-beneath spring'. A related name is **TSAY-YAH-KIN.**

TSAY-YAH-KIN (AZ, Navajo Co.) \TSAY yah kin\. From Navajo *tséyaa kin* 'house under the rock', containing *kin* 'house, building'. A related name is **TSAYATOH.**

TSE BINAAYOLI (NM, McKinley Co.) From Navajo *tsé binááyołí* 'rock (that) wind blows around', containing *tsé* 'rock'.

TSE BITA'I (NM, San Juan Co.) From Navajo *tsé bit'a'í* 'winged rock', containing *tsé* 'rock' and *bit'a'* 'its wings'.

TSE BIYI (AZ, Navajo Co.) \say bi YAH\. From Navajo *tsé biyi* 'rock canyon', containing *tsé* 'rock'. A nearby placename is **Tse Biyi Yazzi**, from *tsé biyi yázhí* 'little rock-canyon'.

TSE BONITA Wash (AZ, Apache Co.; NM, McKinley Co.) A variant spelling of **TSE BONITO.**

TSE BONITO Wash (AZ, Apache Co.; NM, McKinley Co.) \say buh NEE tuh\. From Navajo *tsé binii' tó* 'rock-face water, water on the face of the rock', containing *tsé* 'rock', *binii'* 'its face' (*bi-* 'its', *-nii'* 'face'), and *tó* 'water'. The placename has been folk-etymologized to suggest Spanish *bonito* 'pretty'.

TSE CHIZZI Wash (AZ, Navajo Co.) \say CHEE zee\. From Navajo *tsé ch'ízhí* 'rough rock', containing *tsé* 'rock'.

TSE LIGAI (AZ, Apache Co.) From Navajo *tsé łigai* 'white rock'. A related name is **TSELIGAIDEEZA Canyon** (AZ, Apache Co.).

TSE TAA Ruins (AZ, Apache Co.) From Navajo *tsé táá'á* 'rock extending into water', containing *tsé* 'rock'.

TSE YA TOE Spring (AZ, Navajo Co.) \tsay YAH toh\. Probably Navajo *tséyaa tó* 'water under rock', from *tsé* 'rock', *yaa* 'under', and *tó* 'water'. A related name is **TSE-YAA-TOHI Wash** (NM, San Juan Co.).

TSE YE HA TSOSI Canyon An alternative spelling of **SEGIHATSOSI**. Also spelled **Tseye Ha Tsosi**.

TSEANAZTI Creek (NM, San Juan Co.) A variant spelling of **SANOSTEE**, from Navajo *tsé ałnáozt'i'í* 'where rocks overlap', containing *tsé* 'rock' and *ałnáozt'ti'* 'they overlap'. Many Navajo placenames begin with the element *tsé* 'rock', which is often pronounced like English *say* and sometimes spelled *tsa-*, as in **TSAI SKIZZI Rock**.

TSE-CLANI-TO Wash (NM, San Juan Co.) From Navajo *tsé łání* 'much water' and *tó* 'water'.

TSEDAATAH Canyon (NM, Apache Co.) From Navajo *tsédáá'tah* 'among the rock edges', containing *tsé* 'rock', *dáá'* 'rim, edge', and *tah* 'among'. A related name is **Tseda Hwidezohi Peak** (AZ, Apache Co.), from Navajo *tsédáá' hwiidzohí* 'the place where marks or lines are carved into the edge of the rock', containing *tsé* 'rock', *dáá'* 'edge, rim', *hwii, ho* 'space, area', and *dzoh* 'draw a line or mark'.

TSEGAHODZANI Canyon (AZ, Apache Co.) From Navajo *tséghá hoodzání* 'rock with a hole in it', containing *tsé* 'rock', *ghá* 'through', and *hoodzą́* 'be a perforated area or space'.

TSEGI Canyon (AZ, Navajo Co.) \tuh SAY gee\. From Navajo *tséyi'* 'canyon', lit. 'inside the rocks', with *tsé* 'rock' and *-yi'* 'inside'. The *Chelly* in **Canyon DE CHELLY** represents the same Navajo word. A related name is **TSEGIHATSOSO**.

TSEGIHATSOSO (AZ, Apache Co.) Navajo for 'tight canyon', from *tséyi'* 'canyon' and *hatsoh* 'be tight'. A related name is **TSEGI Canyon**.

TSEGITO Spring (AZ, Apache Co.) From Navajo *tsiyi' tóhí* 'spring in the forest', containing *tsiyi'* 'forest' and *tóhí* 'spring'. A variant spellling is **Segetoa**.

TSEHSO (NM, San Juan Co.) Perhaps from Navajo *tsétsoh* 'big rock', from *tsé* 'rock' and *tsoh* 'big', or perhaps from Kiowa *Xé:xó:* 'thick stone'.

TSEH-YA-KIN Canyon (AZ, Apache Co.) \tsay yah kin\. From Navajo *tséyaa kin* 'house under the rock', with *tsé* 'rock', *yaa* 'beneath', and *kin* 'house'.

TSE'II'AHI (NM, McKinley Co.) From Navajo *tsé íi'áhí* 'rock spire', containing *tsé* 'rock' and *íí'á* 'it stands up, sticks up'.

TSELIGAIDEEZA Canyon (AZ, Apache Co.) From Navajo 'white rock extends', containing *tsé* 'rock', *łigai* 'white', and *deez'á* 'it extends'. A related name is **TSE LIGAI** (AZ, Apache Co.).

TSE-NAS-CHII Wash (NM, San Juan Co.) From Navajo *tsé náshchii'*, containing *náshchii'* 'red in a circle, red ring'.

TSE-NI-CHA Pillar (NM, San Juan Co.) From Navajo *tsé nitsaa* 'big rock', containing *tsé* 'rock'.

TSE-YAA-TOHI Wash (NM, San Juan Co.) From Navajo *tséyaatóhí* 'where there is water beneath the rocks', containing *tsé* 'rock', *yaa* 'beneath', and *tó* 'water'. A related name is **TSE YA TOE Spring** (AZ, Navajo Co.).

TSEYAH (AZ, Apache Co.) Probably from Navajo *tséyaa* 'beneath the rocks', from *tsé* 'rock' and *yaa* 'beneath'.

TSIN KLETZIN (NM, San Juan Co.) \tsin KLET sin\. From Navajo *tsinłizhin* 'charcoal', from *tsin* 'wood, tree' and *łizhin* 'black'.

TSIN LANI Creek (AZ, Apache Co.) From Navajo *tsín łání* 'many trees', containing *tsín* 'wood, tree'.

TSIN SIKAAD (AZ, Apache Co.) From Navajo *tsín sikaad* 'clumps of trees', containing *tsín* 'wood, tree' and *sikaad* 'they sit in clumps'.

TSINAT Mesa (NM, Santa Fe Co.) Probably from Keresan *tsínāuta,* a form of the verb meaning 'to finish, to come to the end', applied to a mountain summit or the edge of a cliff.

TSIN-NAS-KID Summit (NM, San Juan Co.) From Navajo *tsin násk'id* 'trees ringed about with hills', from *tsin* 'tree(s)', *ná(s)* 'encircling, around', and *-k'id* 'hill, hump, mound'.

TSIPING (NM, Rio Arriba Co.) From Tewa *tsip'įŋy* 'Pedernal Mountain', containing *tsi'i* 'flaking stone, obsidian' and *p'įŋy* 'mountain'.

TSISNAATEEL (NM, San Juan Co.) From Navajo *tsis naateel* 'wide hill extending downward', containing *sis, tsis* 'hill, mountain' and *naateel* 'it is wide downward'.

TSITAH Wash (AZ, Apache Co.) From Navajo *tsiitah* 'among the hair', containing *tsii'* 'hair' and *tah* 'among'.

TSO-DZIL Ranch (NM, McKinley Co.) \TSOH tsil\. From Navajo *tsoodził* 'tongue mountain', from *(a)tsoo'* 'tongue' and *dził* 'mountain'. This word has nothing to do with *Tzozil,* the name of a Mayan people and language that is pronounced the same way in English.

TSOSIE Well (AZ, Navajo Co.) \SOH see\. From Navajo *ts'oozí* 'long-haired'; frequently used as a family name.

TSUN-JE-ZHIN (NM, San Juan Co.) From Navajo *tsin ch'ézhin* 'trees extending out black', from *tsin* 'tree', *ch'é-, ch'í-* 'extending out horizontally', and *(łi-) zhin* 'black'.

TUBA City (AZ, Coconino Co.) \TOO buh\. From Hopi *tuuvi,* the nickname of Qötswayma, the founder of Tuba City. It is possibly a shortened form of an alternative personal name, *tuuvi'yma* 'go to throw out', containing *tuuv* 'throw'.

TUBAC (AZ, Santa Cruz Co.) \too BAK, TOO bak\. A Spanish spelling corresponding to Tohono O'odham (Papago) *cewagĭ* 'cloud'. The name was recorded as *Tubac* in the eighteenth century, based on the pronunciation at that time; subsequently, O'odham *t* was replaced by *c* (pronounced *ch*) before high vowels such as *u,* as also in the placename **TUCSON.**

TUCKABATCHEE Church (OK, Hughes Co.) In the historic Creek Nation; perhaps from the name of a Muskogee tribal town, *tukvpvcce.*

TUCSON (AZ, Pima Co.) \TOO sahn\. Spanish *Tucsón,* from Tohono O'odham (Papago) *cukṣon* 'black base', containing *cuk* 'black' and *ṣon* 'base'. The spelling with *t* represents the old pronunciation; since the eighteenth century the pronunciation of *t* before high vowels such as *u* has changed to *ch,* spelled with *c*; see **TUBAC.** There is also a **Tucson Mountain** in New Mexico (Lincoln Co.). A related name on the Tohono O'odham reservation is **ALI CHUKSON.**

TUCUMCARI (NM, Quay Co.) \TOO kum kar ee\. Probably from Comanche *tʉkamʉkarʉ* 'to lie in wait for someone or something to approach'.

TUDECOZ Spring (AZ, Apache Co.) Probably from the same origin as **TOH DE COZ** (AZ, Navajo Co.).

TUKLO Creek (OK, Bryan Co.) \TOO kloh\. In the historic Choctaw Nation; from Choctaw *toklo* 'two'.

TUL, Laguna del (NM, Guadalupe Co.) \luh GOO nuh del TOOL\. The Spanish name means 'lake of the tule or cattail reed', from Spanish *tul, tule* 'cattail reed', from Aztec *tollin.*

TULAPAI Creek (AZ, Gila Co.) \too luh PIE\. Said to refer, in a unidentified Indian language, to a kind of alcoholic drink made from agave cactus.

TULAR, El (TX, Cameron Co.) \too LAHR\. Mexican Spanish for 'a patch of tules or cattail reeds', from *tule* 'cattail reed, tule', from Aztec *tolli* (see **TULE**). A probably related name is **Tolar** (NM, Roosevelt Co.).

TULAROSA (NM, Otero Co.) \too luh ROH suh\. The feminine form of a Spanish adjective meaning 'reedy, having many patches of tules or cattail reeds'. The feminine form of the adjective may come from a phrase such as *tierra tularosa* 'reedy land'. Ultimately from Aztec *tolli.* See **TOLAR, EL TULAR,** and **TULE.**

TULE (TX, Briscoe Co.) \TOO lay, TOO lee\. From Mexican Spanish *tule* 'cattail reed', from Aztec *tolli.* There is also a **Tule** in New Mexico (Eddy Co.).

TULIA (TX, Swisher Co.) \TOOL yuh\. An adaptation of **TULE.**

TULLAHASSEE (OK, Wagoner Co.) \tul uh HAS ee\. In the historic Creek Nation; from Muskogee *tvlvhasse,* the name of a tribal town. The meaning may be 'old town', with *etvlwv* 'tribal town' and *vhassē* 'rancid, stale, old'. Related placenames are *Tallahassee* in Florida (Leon Co.) and perhaps **TULSA.**

TULOSA (NM, San Miguel Co.) \too LOH sah\. This Spanish adjective means 'reedy, full of tules or cattail reeds'; it is derived from *tule* 'cattail reed', from Aztec *tolli*. The feminine form may mean that the name was abbreviated from a phrase such as *tierra tulosa* 'reedy land'. A possible related form is **Tuloso** \too LOH soh\ (NM, San Miguel Co.), apparently the masculine form of the adjective.

TULSA (OK, Tulsa Co.) \TUL suh, TUL see\. County seat of **Tulsa** County, Oklahoma. Perhaps a shortening of Muskogee *lucv-pokv tulse*, the name of a tribal town in Alabama, which may mean 'turtles-sitting old-town', from *lucv-pokv* 'turtles sitting' and *tvlvhassē* 'old town'. There was also a **Tulsa** in Texas (Winkler, Co.), now a ghost town. See **TULLAHASSEE**.

TULSITA (TX, Bee Co.) Named Tulsa by J. L. Courtney after **TULSA**, Oklahoma. Since a **Tulsa** already existed in Texas (now a ghost town in Winkler Co.), the post office in the 1930s was renamed **Tulsita,** a Spanish-style diminutive of **Tulsa**.

TUMACACORI (AZ, Santa Cruz Co.) \too muh KAK uh ree, too muh KAH kuh ree\. Perhaps from Tohono O'odham (Papago) *cemagĭ gakolig* 'caliche bending', where *caliche* refers to 'alkali' and *gakodk* means 'to be bent'. Or perhaps from Tohono O'odham *cem ko'okolig* 'place of little chili', from *cem* 'insufficient' and *ko'okol* 'chili'.

TUMAMOC Hill (Ariz, Pima Co.) \too muh MOK\. From Tohono O'odham (Papago) *cemamagĭ* 'horned lizard'.

TUNAS County (TX, Pecos Co.) \TOO nus\. From Mexican Spanish *tuna* 'fruit of the prickly pear cactus', orginally from the Taino language of the West Indies. The placename **Arroyo las Tunas** (NM, Rio Arriba Co.) is Spanish for 'creek (of) the prickly-pears'.

TUNICHA Mountains (AZ, Apache Co.; NM, San Juan Co.) A variant spelling of **TUNITCHA** and **TUNTSA**.

TUNITCHA Mountains (AZ, Apache Co.; NM, San Juan Co.) \TOO ni chah, too ni CHAH\. From Navajo *tónitsaa'*, 'big water', from *tó* 'water' and *nitsaa* 'big'. Variant spellings are **Tunicha** and **Tuntsa**.

TUNQUE Pueblo (NM, Sandoval Co.) A variant writing of **TONQUE**.

TUNTSA Creek (AZ, Apache Co.) A variant spelling of **Tunicha** and **TUNITCHA**. The name **Tuntsa** also occurs in New Mexico (San Juan Co.).

TUPELO (OK, Coal Co.) \TOO) puh loh\. The term refers to a tree called the *tupelo gum,* perhaps from Muskogee *eto* 'tree' and *opelwv* 'swamp'.

TUSAS (NM, Rio Arriba Co.) \TOO sus\. From Mexican Spanish *tuzas* 'woodrats, prairie dogs', from singular *tuza,* corresponding to Aztec *tuzan* 'wood-rat'. There is also a placename **Las Tusas** (NM, San Miguel Co.), meaning 'the prairie dogs'.

TUSAYAN Ruin (AZ, Coconino Co.) \TOO sah yahn, too sah YAHN, TUS uh yan\. From Spanish *Tusayán,* used by early explorers to designate the area of the Hopi pueblos. Possibly from *tuza* 'wood rat', corresponding to Aztec *tuzan* 'wood rat'. This site, adjacent to Grand Canyon National Park, is a prehistoric pueblo occupied in 1185–1225 by people believed to be ancestors of the Hopis.

TUSCOLA (TX, Taylor Co.) \tus KOH luh\. Probably a transfer from Illinois, Michigan, or Mississippi. In origin it may be an abbreviation of the name of *Tuscolameta Creek* in Mississippi, itself derived from Choctaw *tashka-himmita* 'young warrior', from *tashka* 'warrior' and *himmita* 'young'.

TUSHKA (OK, Atoka Co.) \TUSH kuh\. In the historic Choctaw Nation; from Choctaw *tashka* 'warrior'.

TUSKAHOMA (OK, Pushmataha Co.) \tus kuh HOH muh\. This town was the historic capital of the Choctaw Nation; its name is from Choctaw *tashka-homma* 'red warrior' (lit. 'warrior-red').

TUSKEGEE (OK, Creek Co.) \tus KEE gee\. In the historic Creek Nation; perhaps a transfer from Macon County, Alabama. This was originally the name of a Muskogee tribal town, *taskēke.* The name was first recorded by Spanish explorers as *Tasquiqui,* from dialectal Koasati *taskiki* 'warriors'.

TUWEEP Valley (AZ, Mohave Co.) \TOO weep\. Probably from Southern Paiute *tutu-uippi* 'whirling canyon', from *tutu* 'to whirl' and *uippi* 'canyon'. A variant spelling is **Toroweap.**

TUXEDO (TX, Jones Co.) \TUK si doh\. Perhaps a transfer from Orange County, New York. Probably from Munsee Delaware, perhaps from *p'tuck-sepo* 'crooked river'. The status of the New York town as a resort gave rise to the use of the term *tuxedo* for a man's formal jacket. The Texas town was called Bonita until 1907.

TUYE Spring (AZ, Apache Co.) From the same origin as **TOYEI.**

TUZIGOOT National Monument (AZ, Yavapai Co.) \too zi GOOT, TOO zi goot\. Said to be Western Apache for 'crooked water', from *tó* 'water' and *-god* 'something bent, a knee'.

TYENDE Creek (AZ, Apache Co.) From Navajo *téé'ndééh* '(animals) fall into the water'. A related name is **Kayenta.**

TYNBITO Spring (NM, McKinley Co.) Perhaps from Navajo *tin bito'* 'the spring of ice', from *tin* 'ice', *bi-* 'its', and *to'* 'water, spring'.

TYUONYI (NM, Sandoval Co.) \CHOHN yee\. From the Cochiti (Keresan) placename *tyo'onye,* possibly but not necessarily related to *tyú'uni* 'pottery'.

U

UHANA'A (NM, Cibola Co.) \yoo HAT nah\. The Zuni name of a sacred spring.

UHS KUG (AZ, Pima Co.) From Tohono O'odham (Papago) *u:s ke:k*, lit. 'stick standing'.

UINKARET Mountains (AZ, Mohave Co.) \OON kuh ret\. From Southern Paiute *yïpinkatïtï* 'Mount Trumbull', lit. 'pine peak', from *yïpim-pï* 'pine tree'and *katïtï* 'peak', from *katï-* 'to sit'.

UNA VIDA, Pueblo (NM, San Juan Co.) \OO nuh VEE duh\. The Spanish name of this ancient Indian village means 'one-life', perhaps referring to a living tree stump growing in the archaeological site.

UNAP Mountain (OK, Kiowa Co.) Based on a Kiowa surname.

UNCAS (OK, Kay Co.) \UNG kus\. The name of a seventeenth-century Mohegan leader, derived from *wonkus* 'fox'; the name was made well known through James Fenimore Cooper's novel *The Last of the Mohicans*.

UNCHUKA (OK, Coal Co.) \un CHOO kuh\. From Choctaw *ą-chokka* 'my house'.

UNKAR Creek (AZ, Coconino Co.) \UN kar\. Perhaps from Southern Paiute, meaning 'red creek', from *anka-* 'red'.

UNSHAGI Ruin (NM, Sandoval Co.) From Jemez (Towa) *unšaagi* 'place surrounded by cedar trees', containing *un* 'cedar', *šaa* 'to surround', and *-gi* 'place'.

UQUALLA Point (AZ, Coconino Co.) \OO kwah luh, UK wah luh\. The surname of a Havasupai family from the area.

UTE \yoot\. The name of a Numic-speaking people (their language is a branch of the Uto-Aztecan family) living in Colorado and Utah; they are now often distinguished as the Northern Utes, in northern Utah, and the Southern Utes, in southern Colorado. The term is from Spanish *yuta,* perhaps a borrowing from Western Apache *yúdah* 'high', meaning 'in the mountains'. The **Ute Mountain Indian Reservation** is in southeastern Colorado and extends into New Mexico (San Juan Co.). There is a **Ute Lake** in New Mexico (Harding and Quay Cos.).

UTEVAK (AZ, Pima Co.) From Tohono O'odham (Papago) *uḍuwhag* 'cattail reed'.

V

VAHKI (AZ) From Pima *wa'akih* 'ancient house'.

VAINOM KUG (AZ, Pima Co.) On the Tohono O'odham (Papago) Reservation; from Tohono O'odham *wainomi ke:k* 'an iron standing', containing *wainomi* 'a flatiron' and *ke:k* 'to stand'.

VAIVA VO (AZ, Maricopa Co.) \VEE vah voh\. On the Tohono O'odham (Papago) Reservation; from Tohono O'odham *waiwa wo'o* 'cocklebur pond'.

VAKAMOK (AZ, Pima Co.) On the Tohono O'odham (Papago) Reservation; from Tohono O'odham *wakimagi* 'to be worn, ragged'.

VAMORI (AZ, Pima Co.) On the Tohono O'odham (Papago) Reservation; from Tohono O'odham *wamel* 'swamp'.

VAYA CHIN (AZ, Pima Co.) On the Tohono O'odham (Papago) Reservation; from Tohono O'odham *wawhia chini* 'well mouth'.

VICKS Peak (NM, Socorro Co.) \viks\. Named for the Apache leader Victorio. See **VICTORIO Mountains.**

VICTORIO Mountains (NM, Luna Co.) \vik TORE ee oh\. This was the Spanish name of a Mimbreño Apache leader. It is said that he was known in Apache as *bidóóya*; the relationship between the Spanish and the Apache forms is unclear. In Texas (Culberson Co.), **Victorio Peak** is also named for the Apache warrior.

VIOPUL (AZ) From Tohono O'odham (Papago) *wiwpul* 'wild tobacco'.

VOPOLO HAVOKA (AZ) From Tohono O'odham (Papago) *wuplu ha-wo'oka* 'donkeys' pond', from *wuplu* 'donkeys' (sing. *wu:lu*) and *wo'o* 'pond'.

W

WABAKWA Ruin (NM, Sandoval Co.) From the Jemez (Towa) placename *waaba-kwa,* containing *-kwa* 'place'.

WACO (TX, McLennan Co.) \WAY koh\. From Wichita *wi:ko,* referring to a tribal subgroup. The term was earlier written *Wico* by Anglo-Americans and *Hueco* by Spaniards. There was a post office at **Waco,** Oklahoma (Pottawatomie Co.) from 1900 to 1904, and present **CHICKASHA** (OK, Grady Co.) was called Waco from 1890 to 1891. Descendants of the Waco tribe now form part of the **Wichita and Affiliated Tribes** at Anadarko, Oklahoma (Caddo Co.). **Waco Lake** is near the town of Waco, Texas.

WADE (OK, Bryan Co.) Named for Alfred Wade, a prominent Choctaw man.

WADENA (OK, Pushmataha Co.) Probably a transfer from Minnesota. Said to have been named for an Ojibwa leader in the early nineteenth century. The word may be a shortened form of a name ending in *-adinaa* 'hill'.

WAHAK HOTRONTK (AZ, Pima Co.) From Tohono O'odham (Papago) *wo:g huḍuñĭk* 'a dip in the road', from *wo:g* 'road' and *huḍuñĭk* 'a dip'.

WAHOO Bay (OK, Wagoner Co.) \WAH hoo\. An inlet in Fort Gibson Lake, on the Neosho River. The name may be a transfer from Georgia (Lumpkin Co.), where it refers to the winged elm, *Ulmus alata,* and is based on Muskogee *vhahwv* 'walnut'. Alternatively, it may be from Nebraska (Saunders Co.), where it refers to a shrub called the burning bush, *Eunonymus atropurpureus,* from Dakota *wą-hu,* lit. 'arrow-wood'. The word was popularized in the 1930s by a song that included the words "Gimme a horse, a great big horse, and gimme a buckaroo / And let me Wah-Hoo! Wah-Hoo! WAH-HOO!" The source of the placename, however, seems to be botanical rather than the exclamation used in the song.

WAH-SHA-SHE State Park (OK, Osage Co.) \wah SHAH she\. From the Osage word for their own tribe, *wažáže.* The park is located on Hulah Lake.

WAHWEAP (AZ, Coconino Co.) \WAH weep\. Perhaps from Southern Paiute *wa'a-uippi* 'cedar canyon', from *wa'a-* 'cedar, pinyon pine' and *uippi* 'canyon'.

WAKA (TX, Ochiltree Co.) \WAH kuh\. Said to be shortened from *Wawaka,* meaning 'wet or swampy ground' in an unidentified Indian language. Another possible source is Comanche *waaka* 'gallop'.

WAKITA (OK, Grant Co.) \wah KEE tuh\. Said to be Cherokee, referring to water collected in a depression such as a buffalo wallow.

WALAPAI (AZ, Mohave Co.) \WAH luh pie\. This placename refers to a tribe of the Yuman family living in northern Arizona; the tribal name is also spelled *Hualapai.* The name is from Mojave *huuwaalyapay* 'pine tree people', from *huwaaly* 'pine tree'. Related forms, all in Arizona, are **HAULAPAI** (Mohave Co.), **HUALAPAI** (Cocoino Co.), and **Hualpai** (Yavapai Co.).

WALKER'S Station (OK, Le Flore Co.) Said to be named for Tandy Walker, a Choctaw leader. The name may be derived from Choctaw *waaka* 'cow', in turn derived from Spanish *vaca*. But Walker is also a common Creek surname, perhaps from Muskogee *wakv* 'cow'.

WALLA Valley (AZ, Coconino Co.) \WAH luh\. From Hopi *waala* 'gap'.

WALPI (AZ, Navajo Co.) \WAHL pie\. From Hopi *wàlpi* 'place of the gap', from *waala* 'gap'.

WAMUL Tank (AZ, Pima Co.) \WAH mul\. From Tohono O'odham (Papago) *wamel* 'swamp'.

WANETA (TX, Houston Co.) \wah NEE tuh\. Supposedly the name of an Indian woman, but the name may be a respelling of the Spanish *Juanita*, familiar from a popular song of the nineteenth century.

WAPANUCKA (OK, Johnston Co.) \wah puh NOO kuh\. Perhaps a transfer name from *Wapanacki Lake* (Vermont, Lamoille Co.) or *Wapanocca Bayou* (Arkansas, Crittenden Co.), both of which come from an Algonquian word meaning 'easterners'. (Consider the related Ojibwa word *waaban* 'east' and the Abenaki word *wo^banaki* 'the east').

WASHINGTON (OK, McClain Co.). Named for George Washington, a Caddo chief.

WASHITA County (OK) \WAH shi tah\. An alternative form of **OUACHITA**, the name of a Caddoan people. The **Washita River** rises in the Texas Panhandle and flows through Oklahoma, emptying into the Red River in Marshall County. The **Washita Battlefield National Historic Site**, near **CHEYENNE**, Oklahoma (Roger Mills Co.), commemorates the U.S. Army's 1868 surprise attack, led by by Lt. Col. George Custer, on the Cheyenne, led by Chief Black Kettle. The **Fort Washita Historic Site,** near Durant, Oklahoma (Bryan Co.), has a museum commemorating the fort built there in 1842 to encourage settlement by the Chickasaw and Choctaw tribes, protecting them from raids by Plains Indians. The **Washita National Wildlife Refuge**, on the Washita River, adjoins Foss Lake in Custer County, Oklahoma.

WASHUNGA (OK, Kay Co.) \wah SHOONG guh\. Named for a leader of the Kansa tribe; his Native name was *wašúǵe*.

WATAHOMIGIE Point (AZ, Coconino Co.) \wah tuh HOH mi jee\. From the surname of a Havasupai family. The Native name is *watahómija*, perhaps meaning 'stripped house', with *wa* 'house' and *tahómiga* 'to strip off'.

WATAUGA (TX, Tarrant Co.) \wuh TA guh, wuh TAW guh\. Probably a transfer name from Tennessee. Earlier written *Watagi, Watogo*, and so forth, perhaps meaning 'village of many springs'. Said to be the name of a Cherokee village, recorded in 1771.

WATONGA (OK, Blaine Co.) \wah TONG guh, wuh TONG guh\. Named for an Arapaho leader whose Native name is said to have been *wó'teen-koo'óh* 'black coyote'. **Watonga Lake** is nearby. Also written *Wa-ton-gha* in early records.

WATOVA (OK, Nowata Co.) \wah TOH vuh\. Named for an Osage leader; the term is said to mean 'spider', but it is doubtful that this is an Osage name.

WAUHILLAU (OK, Adair Co.) Said to be from Cherokee *awa'hili* 'eagle'.

WAUKEGAN (TX, Montgomery Co.) Transferred from Illinois (Lake Co.); perhaps from Potawatomi *wakāígín* 'fort' or Ojibwa *wâkaigan* 'fort'.

WAUKOMIS (OK, Garfield Co.) \wah KOH mis, waw KOH mis\. Perhaps from an unidentified Indian language.

WAURIKA (OK, Jefferson Co.) \wah REE kuh, waw REE kuh\. Perhaps from Comanche *wo'arɨhka,* the name of a tribal band, meaning literally 'worm eaters', from *wo'a* 'worm' and *tɨhka* 'eat'. Near to the town are **Waurika Lake** and **Waurika Wildlife Management Area.**

WAXAHACHIE (TX, Ellis Co.) \wahk suh HACH ee\. Perhaps a transfer name from *Waxahatchee*, Alabama (Shelby Co.), which is perhaps from Muskogee *wakse*, a clan name, and *hvcce* 'stream'. Or perhaps from something close to the Wichita (Caddoan) *waks'ashe:c'a* 'fat monster', containing *waks* 'a mythical monster' and *he:c* 'fat'. **Waxahachie Creek** and **Lake Waxahachie** are nearby.

WAXHOMA, Lake (OK, Osage Co.) \waks HO muh\. Possibly a blend containing -*homa*, from **OKLAHOMA.**

WAYNOKA (OK, Woods Co.) \way NOH kuh, wuh NOH kuh\. Said to be from Cheyenne *tseve'eeno'e* 'that which tastes sweet', containing *vé(k)e-* 'sweet' and -*éno'e* 'taste'.

WEALAKA (OK, Tulsa Co.). From Muskogee *ue-vlvkv*, lit. 'water-rising'. The town is at the site of **Wealaka Mission**, established in 1881.

WEBBER'S FALLS (OK, Muskogee Co.). Named for Walter Webber, a nineteenth-century Cherokee chief.

WECHARTY (OK, Hughes Co.) \wuh CHAR tee\. Said to be named after a blacksmith shop, reflecting Muskogee *wecattetv* 'to make sparks'.

WECHES (TX, Houston Co.) \WEE chiz\. Said to have been adapted from the name of the nearby **Neches River**. See NECHES.

WEESATCHE (TX, Goliad Co.) \WEE sach\. From Mexican Spanish *huisache* 'acacia (a tree or shrub)', from Aztec *huixachin.*

WEKIWA (OK, Tulsa Co.) \wuh KEE wuh\. From Muskogee *uekiwv* 'a spring'.

WELEETKA (OK, Okfuskee Co.) \wuh LEET kuh\. From Muskogee *ueletkv*, lit. 'water running,' containing *ue-* 'water'. **Weleetka Lake** is nearby.

WEPO Village (AZ Navajo Co.) \WEE poh, WAY poh\. From Hopi *wipho* 'cattail reed'.

WETUMKA (OK, Hughes Co.) \wee TUM kuh\. From Muskogee *wetumhkv*, perhaps meaning 'tumbling water', from *ue-* 'water'. **Lake Wetumka** is nearby.

WEWOKA (OK, seat of Seminole Co.) Historic capital of the Seminole Nation; from Muskogee *ue-wohkv* 'barking water, roaring water', containing *ue* 'water'. Near the town is **Wewoka Lake.**

WHEELING (OK, Comanche Co.) \HWEE ling\. Probably a transfer from West Virginia. Said to be from Delaware (Algonquian), meaning 'at the head' or 'on the head', from *wil* 'head' and *ing* 'at' (locative ending), perhaps referring to a decapitation.

WHERE THE YEI WENT UP (AZ, Apache Co.) A translation of Navajo *yee'ii hadeeyáhí* 'where the yei went up and out'; the word *yei* refers to supernatural beings.

WHIRLWIND Cave (AZ, Apache Co.) Corresponds to Navajo *tséyaa hahwiiyoolí* 'where the wind eddies under the rock'.

WHISKEY Creek (AZ, Apache Co.) Corresponds to Navajo *tó dithił bikooh,* from *tó dithił* 'whiskey' and *bi-kooh* 'its creek'. But *tó dithił* probably meant 'dark water' originally; *dithił* means both 'causes dizziness' and 'is dark-colored'. A related placename is **TO-DIL-HIL Wash** (NM, San Juan Co.).

WHITE BEAD (OK, Garvin Co.) Named for a Caddo leader known as Chief White Bead.

WHITE CLAY Wash (AZ, Apache Co.) From Navajo *dleesh łigai,* lit. 'clay-white'.

WHITE EAGLE (OK, Kay Co.) Named for a Ponca leader. Also spelled **Whiteagle.**

WICHITA (OK, Comanche Co.) \WICH i tah, WICH i taw\. The placename is that of a people and language of the Caddoan linguistic family, living in the southern Great Plains. The term *Wichita,* earlier written **OUACHITA,** *Ousita,* and so forth, was originally the name of a single village group, but it came to be applied to a group of related peoples, including the Wacos and the Kitsais, or Keechis. In the twentieth century, the Wichitas came to call themselves *kirikir'i:s* 'raccoon-eye(s)', which was also originally the name of a single band. The headquarters of the **Wichita and Affiliated Tribes** is in Anadarko, Oklahoma (Caddo Co.); the affiliated tribes are the **Waco, Tawakoni**, and **Keechi**. Oklahoma also has the **Wichita Mountains** (Comanche Co.). Texas contains the placename **Wichita County** and the county seat of **Wichita Falls;** the **Wichita River** runs into the Red River nearby.

WIDE RUINS (AZ, Apache Co.) Corresponds to Navajo *kinteel* 'wide house', from *kin* 'house' and *-teel* 'wide'. A related name is **KIN TEEL** (NM, McKinley Co.).

WIJIJI (NM, San Juan Co.) \wi JEE jee\. From Navajo *diwózhiishzhiin* 'black greasewood', containing *diwózhii* 'greasewood' and *-(sh)zhiin* 'black'.

WILD CHERRY Canyon (AZ, Apache Co.) Corresponds to Navajo *didzé sikaad* 'chokecherry (bush) sits', containing *didzé* 'chokecherry'.

WILDCAT Trading Post (NM, McKinley Co.) Corresponds to Navajo *náshdói ba'áan* 'wildcat's cave', from *náshdói* 'wildcat' and *ba'áán* 'its cave' (*ba-, bi-* 'its', *a'áan* 'cave').

WILLAHA (AZ, Coconino Co.) \wi LAH hah\. Said to be Havasupai for 'watering place'.

WILL ROGERS Birthplace Ranch (OK, Rogers Co.) In the traditional Cherokee Nation, near **OOLOGAH**, birthplace of the Cherokee writer, humorist, and film personality Will Rogers (1879–1935). The **Will Rogers Memorial Museum** is in Claremore (also in Rogers Co.).

WINDOW Rock (AZ, Apache Co.) Corresponds to Navajo *tségháhoodzání* 'rock with a hole through it', containing *tsé* 'rock', *ghá* 'through', and *hoodzą́* 'perforated area or space'.

WINDSWEPT TERRACE (AZ, Navajo Co.) Corresponds to Hopi *huk'ovi* 'high windy place', from *huk* 'wind', *'o* 'high', and *-vi* 'place'.

WINDY Canyon (NM, McKinley Co.) Corresponds to Navajo *tsé bii'naayolí* 'wind blowing around in the rocks', containing *tsé* 'rock', *bii'* 'in it/them', and *naayolí* 'wind blowing around'. The same Navajo name probably underlies the placename **Windy Rock** (AZ, Apache Co.).

WINGVA (AZ, Navajo Co.) Earlier *Wein-Bahi*; perhaps from Hopi.

WINONA County (OK, Smith Co.) \wi NOH nuh\. A transfer name from the upper Midwest, perhaps from Minnesota; derived from Dakota *winóna* 'first-born child, if a daughter'. The name was that of Hiawatha's mother in Longfellow's poem, and it became immensely popular in the nineteenth century both as a woman's given name and as a placename. The placename **Winona** also occurs in Texas (Smith Co.).

WITCH WATER POCKET (AZ, Mohave Co.) Perhaps a translation of Southern Paiute *innupin picabu* (perhaps related to *inippi-* 'evil spirit, devil, ghost').

WODO, Mount (AZ, Coconino Co.) \WOH doh\. This is the surname of a Havasupai family.

WUKOKI Ruin (AZ, Coconino Co.) \woo KOH kee, wuh KOH kee\. The name is apparently composed of Hopi elements, supposedly meaning 'big house'; *wuko-* means 'big', and *-ki, kii-hu* means 'house'.

WUKOPAKABI (AZ, Apache Co.) \wuh KOH puh kah bee\. From Hopi *wukovaqavi* 'big reed place', from *wuko* 'big', *vaqa, paaqa* 'reed', and *vi* 'place'.

WUPATKI National Monument (AZ, Coconino Co.) \woo PAHT kee, wuh PAHT kee\. From Hopi *wupatki* 'long cut', containing *wupa-* 'long' and *-tk-, -tuku* 'cut', referring to a wash.

WYANDOTTE (OK, Ottawa Co.) \WIE un dot, WINE dot\. Named for an Iroquoian people and language, originally closely associated with the Hurons; their self-designation was *wẹdat,* perhaps a shortening of a longer form corresponding to Mohawk *skawẹ:nat* 'one language' or *tsha'tekawẹnat* 'the same language'. Alternatively, the name may be related to Huron *wendat* 'forest' and/or *yandata* 'village'. The town of Wyandotte, Oklahoma, is headquarters of the **Wyandotte Nation.**

WYNONA (OK, Osage Co.) \wie NOH nuh\. Probably adapted from **WINONA.**

Y

YAHOLA (OK, Muskogee Co.) \yuh HOH luh\. Said to be named for a
Muskogee Indian named Yahola Harjo; from a Creek family name, *Yvholv*,
originally a title appearing in war names and thought to refer to one
responsible for giving a special whoop at the Green Corn Ceremony.
A possible alternative derivation is as a transfer name from Yahoola
Creek, Georgia (Lumpkin Co.), which in turn has been said to be from
a Cherokee personal name, perhaps based on *yahula* 'doodle bug'. **Lake
Yahola** is a reservoir in Tulsa County, Oklahoma.

YAH-TA-HEY (NM, McKinley Co.) \YAH tuh hay\. From the Navajo greeting
yá'át'ééh, lit. 'it is good'.

YAKI Point (AZ, Coconino Co.) \YAH kee, YAK ee\. Named for the Yaqui
Indians of Sonora, Mexico, many of whom now live in Arizona. The term
is from their name for themselves, *hiaki.* Coconino County also has a
Yaqui Crater, from the same source.

YAMPAI (AZ, Yavapai Co.) \YAHM pie, YAHM pee\. The name was earlier
recorded as both *Yampa* and *Yampai.* Thus the word may be related to
Yampa, a placename in Colorado (Routt Co.), from Ute *yampa* 'an edible
root', or it may be from *Yavapai,* the name of a Yuman people.

YANKEE Bill Prairie (OK, Ottawa Co.) This term for an Anglo-American,
especially one from New England, is perhaps from Algonquian *yengees,*
adapted from the word *English.* However, there may be influence from
Cherokee *eankke* 'slave, coward, prisoner', the modern form of which
is *ayvgi* 'captive'. A derivation from a Dutch nickname, *Jan Kees* 'John
Cheese', is also possible. The term *Yankee* occurs as a personal nickname
in placenames such as **Yank's Canyon** (AZ, Santa Cruz Co.).

YANUBBE River (OK, McCurtain Co.) Flows into the Little River. Perhaps
related to **YARNABY**.

YANUSH (OK, Latimer Co.) \YAH nush\. From Choctaw *yanash* 'buffalo'.

YAPASHI (NM, Sandoval Co.) \yuh PAH shee\. A shortening of Cochiti
yapašenye 'sacred enclosure'.

YAPONCHA (AZ, Coconino Co.) From Hopi *yaapontsa,* the name of a
supernatural being associated with windstorms.

YAQUI Crater (AZ, Coconino Co.) \YAH kee, YAK ee\. Named for the Yaqui
Indians of Sonora, Mexico, many of whom now live in Arizona. The term
is from the self-designation *hiaki.* The **Pascua Yaqui Indian Reservation**
is south of the city of Tucson (Pima Co.). Coconino County also has a
Yaki Point, from the same source.

YARNABY (OK, Bryan Co.) \YAR nuh bee, YAH nuh bee\. Perhaps from the war name of a Choctaw man, meaning 'to go and kill', containing *iya* 'to go' and *abi* 'to kill'. Perhaps related to **YANUBBE River.**

YASHAU Creek (OK, McCurtain Co.) \YAH shau\. Perhaps from Choctaw *iyaasha* 'large kettle used for cooking hominy with pork'. Yashau Creek flows into the Little River.

YAVA (AZ, Yavapai Co.) \YAV uh, YAH vuh\. Said to be an abbreviation of *Yavapai.*

YAVAPAI County (AZ) \YAH vuh pie, YAV uh pie\. The name refers to a Yuman people; the word is the plural of *yavpé,* the name for a member of the largest tribal subgroup.

YAZZI Well (NM, San Juan Co.) \YAH zee, YAZ ee\. From Navajo *yázhí* 'small (person)'. This is a common Navajo surname, usually spelled *Yazzie* in English. The placename **Yazzi** also occurs in Arizona (Apache Co.). An adaptation of the same word is **Yazzie Mesa** (AZ, Apache Co.).

YEI BICHEI (AZ, Navajo Co.) \YAY bi chay\. From Navajo *ye'ii bicheii* 'grandfather of the ye'ii', referring to a type of supernatural being.

YELLOW JACKET Canyon (AZ, Coconino Co.) A translation of Navajo *tsís'náłtsooí* 'yellowjacket, a kind of hornet', from *tsís'na* 'bee, wasp' and *-łtso-* 'yellow'.

YON DOT Mountains (AZ, Coconino Co.) From Navajo *yaa ndee'nil* 'series of hills going down'.

YONKIPIN Lake (OK, Rogers Co.) Perhaps from Choctaw.

YSLETA (TX, El Paso Co.) \is LET uh\. This is a spelling variant of Spanish **ISLETA** 'little island', as applied to a Tiwa pueblo in New Mexico (Bernalillo Co.), from which some Indians moved to the El Paso area. The Texas site is also called **Ysleta del Sur** and **TIGUA Pueblo.**

YSLETAÑO Canyon (NM, Otero Co.) The Spanish word means 'belonging to Ysleta', referring to **ISLETA Pueblo** (NM, Bernalillo Co.).

YUBA (OK, Bryan Co.) \YOO buh\. A transfer from Yuba County, California. From the name of a Maidu village, earlier written *Yubu, Yupu,* and *Jubu;* from Nisenan *yubuy* 'shade, shadow'.

YUCHI \YOO chee\. The name of a Native people politically assocated with the Muskogees but speaking a separate language. They originally lived in Georgia and Alabama, but their descendants now form the **Yuchi (Euchee) Tribe**, a division of the Muscogee Creek Nation living at Sapulpa, Oklahoma (Creek Co.). The Muskogee form of the name is *yocce.* A related placename is **EUCHA** (OK, Delaware Co.).

YUGEEHA YUSTEN (OK, Creek Co.). \YU jee ha yus ten\. The name of a section of Mission Boulevard that runs in front of the high school on the site of the old Euchee Indian mission in Sapulpa, Oklahoma. It means

"Euchee (Yuchi) Street," spelled *yUdjEha yust'ê* in the spelling of the Euchee tribal language project.

YUKON (OK, Canadian Co.) \YOO kon\. A transfer from the name of the river in Canada and Alaska. From an Athabaskan language; perhaps Koyukon *yookkene* or Lower Tanana *yookuna*.

YUMA County (AZ) \YOO muh\. From the name of an Indian people of the Yuman language family. The term is probably from Tohono O'odham (Papago) *yu'mĭ*, the name that the Piman peoples applied to the Yuma people. In recent years, the Yumas have restored their traditional self-designation, *Quechan* or *Kwtsaan* \kwuh CHAHN\.

YUMTHESKA Mesa (AZ, Coconino Co.) The surname of a Havasupai family.

YUNOSI Point (AZ, Coconino Co.) The surname of a Havasupai family.

YUPON (TX, Colorado Co.) \YOO pon\. Perhaps transferred from the Carolinas. The word refers to a plant, also called *yupon* or *yupon holly*, from Catawba *'yą́pą'*.

Z

ZACAHUISTLE Pasture (TX, Brooks Co.) \sah kah WEEST lay\. Mexican Spanish, referring to a native plant; from Aztec *zacahuitztli,* containing *zacatl* 'grass' and *huitztli* 'thorn'. Related placenames are **ZACAWEISTA Ranch** (TX, Wilbarger Co.) and **SACAHUISTE Draw** (NM, Eddy Co.).

ZACATA Creek (TX, Webb Co.) \zuh KAH tay\. From Mexican Spanish *zacate* 'grass, hay, fodder', pronounced approximately \sah KAH tay\; from Aztec *zacatl* 'grass'. *Zacate* itself is not currently used as a placename, but it occurs in the names **SACATE, ZACATES,** and **Arroyo ZACATOSO.**

ZACATES (NM, Bernalillo Co.) The plural of Mexican Spanish *zacate* 'grass'; see **ZACATA Creek.**

ZACATOSO, Arroyo (TX, Zapata Co.) Spanish for 'grassy creek', from Mexican Spanish *zacate* 'grass, hay'; see **ZACATA Creek.** A related placename is **SACATOSA Mesa.**

ZACAWEISTA Ranch (TX, Wilbarger Co.) \ZAK uh wis tuh\. Derived from **ZACAHUISTLE Pasture** (TX, Brooks Co.).

ZIA Pueblo (NM, Sandoval Co.) \ZEE uh\. From Keresan *ts'îiya,* the Native name. Also written **Sia.**

ZILBETOD Peak (AZ, Apache Co.) \zil bay TOHD\. From Navajo *dził béét'óód* 'bald mountain' or *dził bét'ood* 'balding mountain', containing *dził* 'mountain'.

ZILDITLOI Mountain (NM, McKinley Co.) \ZIL di tloi\. From Navajo *dził ditłoii* 'fuzzy mountain', containing *dził* 'mountain' and *ditł'o* 'it is fuzzy'.

ZILLESA Mesa (AZ, Navajo Co.) From Navajo *dził deez'á* 'where the mountain range begins', containing *dził* 'mountain' and *deez'á* 'it extends'.

ZILNEZ Mesa (AZ, Navajo Co.) \zil NEZ\. From Navajo *dził ninééz* 'long mountain', containing *dził* 'mountain'.

ZILTAHJINI Peak (AZ, Navajo Co.) From Navajo, meaning 'standing cranes', containing *dééł* 'a crane (bird)'.

ZUNI Pueblo (NM, McKinley Co.) \ZOO nee\. The name refers to the Indian people and language that are native to the pueblo; from Spanish *ʒuñi,* from a Keresan name like Acoma *sî:ni* 'Zuni Pueblo'. The placename also occurs in Arizona (Navajo Co.).

ZUZAX (NM, Bernalillo Co.) \ZOO zaks\. A name invented by a curio shop owner, intended to attract attention but attributed by him to the nonexistent "Zuzax Indians".

Appendix 1

LANGUAGES AND LANGUAGE FAMILIES
MENTIONED IN THIS GUIDE

The languages that make up language families, like members of biological families (genera), show shared characteristics and are assumed to be descended from a common ancestor. In linguistics, the ancestral language is called a *proto-language*—for example, Proto-Uto-Aztecan and Proto-Indo-European. The following list of Native languages mentioned in this book shows to which language family each belongs, and the subsequent list tells which related languages are subsumed under each family. Languages labeled "Isolate" have no proven relationships to other languages.

Language	*Family*
Acoma	Keresan
Akimel O'odham (Pima)	Uto-Aztecan
Alabama	Muskogean
Aleut	Eskimo-Aleut
Apache (Chiricahua, Plains, Lipan, etc.)	Athabaskan
Arapaho	Algonquian
Aymara	Aymaran
Aztec (Nahua, Nahuatl, Mexica, Mexicano)	Uto-Aztecan
Caddo	Caddoan
Catawba	Siouan
Cayuga	Iroquoian
Cherokee	Iroquoian
Cheyenne	Algonquian
Chickasaw	Muskogean
Choctaw	Muskogean
Cochiti (Kochiti)	Keresan
Cocopah	Yuman
Comanche	Uto-Aztecan
Coushatta (Koasati)	Muskogean
Creek (Muskogee, Muscogee)	Muskogean
Crow	Siouan
Dakota	Siouan
Delaware (Lenape)	Algonquian

Language	Family
Diegueño	Yuman
Euchee (Yuchi)	Isolate
Gabrielino	Uto-Aztecan
Gwich'in	Athabascan
Havasupai	Yuman
Hitchiti (Mikasuki)	Muskogean
Hopi	Uto-Aztecan
Hualapai (Walapai)	Yuman
Huastec Mayan	Mayan
Illinois (Miami)	Algonquian
Iowa (Ioway)	Siouan
Karankawa (an extinct language)	Isolate
Keechi (Kitsai)	Caddoan
Kewa Pueblo (formerly Santo Domingo)	Keresan
Kiowa	Kiowa-Tanoan
Kitsai (Keechi)	Caddoan
Klamath	Penutian
Koasati (Coushatta)	Muskogean
Koyukon	Athabascan
Kwtsaan (Quechan, Yuman)	Yuman
Lakota (Lakhota)	Siouan
Lenape (Delaware)	Algonquian
Lower Tanana	Athabascan
Mahican	Algonquian
Maidu	Penutian
Maricopa	Yuman
Mayan (Huastec, Jacaltec, Yucatec, etc.)	Mayan
Miami (Illinois)	Algonquian
Mikasuki (Hitchiti)	Muskogean
Modoc	Penutian
Mohawk	Iroquoian
Mohegan	Algonquian
Mojave (Mohave)	Yuman
Muskogee (Creek, Muscogee)	Muskogean
Nahua (Aztec, Mexica, Mexicano, Nahuatl)	Uto-Aztecan
Natchez	Isolate
Navajo (formerly Navaho)	Athabaskan
Nevome	Uto-Aztecan
Nisenan	Penutian

Language	Family
Ojibwa (Ojibway)	Algonquian
Oneida	Iroquoian
Onondaga	Iroquoian
O'odham (Pima, Papago, Nevome)	Uto-Aztecan
Ópata	Uto-Aztecan
Osage	Siouan
Otoe	Siouan
Ottawa	Algonquian
Papago (Tohono O'odham)	Uto-Aztecan
Pawnee	Caddoan
Pecos	Kiowa-Tanoan
Pima (Akimel O'odham)	Uto-Aztecan
Piro	Keresan
Ponca	Siouan
Pottawatomi (Potawatomi)	Algonquian
Quapaw	Siouan
Quechan (Kwtsaan, Yuma)	Yuman
Quechua	Quechuan
Sac (Sauk)	Algonquian
Sahaptin	Sahaptian
Sauk (Sac)	Algonquian
Seminole	Muskogean
Seneca	Iroquoian
Shawnee	Algonquian
Southern Paiute	Uto-Aztecan
Taino	Arawakan
Tarahumara	Uto-Aztecan
Tewa	Kiowa-Tanoan
Tiwa	Kiowa-Tanoan
Tohono O'odham (Papago)	Uto-Aztecan
Tompiro (an extinct language)	Unknown
Tonkawa	Isolate
Towa (Jemez Pueblo)	Kiowa-Tanoan
Ute	Uto-Aztecan
Virginia Algonquian	Algonquian
Waco	Caddoan
Walapai (Hualapai)	Yuman
Washoe (Washo)	Isolate
Wichita	Caddoan

Language	Family
Wyandotte	Iroquoian
Yaqui	Uto-Aztecan
Yavapai	Yuman
Yuchi (Euchee)	Isolate
Yuma (Kwtsaan, Quechan)	Yuman
Zuni	Isolate

Language Families Mentioned in This Guide, with Some of Their Better-Known Languages

Algonquian	Arapaho, Blackfoot, Cheyenne, Cree, Delaware (Lenape), Fox, Illinois, Miami, Ojibwa, Ottawa, Potawatomi, Sauk, Shawnee
Arawakan	Taino and languages of South America and the Caribbean
Athabaskan	Apache languages, Navajo
Aymaran	Aymara
Caddoan	Caddo, Keechai (Kitsai), Pawnee, Wichita
Eskimo-Aleut	Aleut, Eskimo
Iroquoian	Cayuga, Cherokee, Mohawk, Oneida, Onondaga, Seneca
Keresan	Pueblo Indian languages spoken at Acoma, Cochiti (Kochiti), and Kewa (Santo Domingo), among other pueblos
Kiowa-Tanoan	Kiowa and the Pueblo Indian languages Tewa, Tiwa, and Towa (Jemez)
Muskogean	Alabama, Apalachee, Chickasaw, Choctaw, Coushatta (Koasati), Creek (Muskogee), Hitchiti (Mikasuki), Seminole
Penutian	Klamath, Maidu, Modoc, Yokuts
Quechuan	Quechua (Inca) and other South American languages
Sahaptian	Nez Perce and others of the southern Columbia Plateau region of the United States
Siouan	Crow, Dakota, Lakhota, Nakhota, Osage, Otoe, Ponca, Quapaw, Catawba
Tanoan	Tewa, Tiwa, Towa
Uto-Aztecan	Aztec, Comanche, Hopi, O'odham (Piman languages), Southern Paiute, Tarahumara, Ute, Yaqui; others of Central and North America
Yuman	Cocopa, Havasupai, Hualapai (Walapai), Maricopa, Mojave (Mohave), Yavapai

TRIBAL CONTACT INFORMATION

The following information was verified with the Arizona Commission of Indian Affairs, the New Mexico Indian Affairs Department, the Oklahoma Commission on Indian Affairs, and individual Texas tribal websites. Please make the University of Oklahoma Press aware of any omissions or updated information. We have attempted to provide addresses for official websites wherever possible; for New Mexico pueblos, some information is available at www.indianpueblo.org.

Arizona

Ak Chin Indian Community Council
    42507 W. Peters and Nell Road, Maricopa, AZ 85239
    (520) 568-2227; www.ak-chin.nsn.us

Cocopa Tribe
    County 15th and Avenue G, Somerton, AZ 85350
    (928) 627-2061; www.cocopah.com

Colorado River Indian Tribes
    26600 Mohave Road, Parker, AZ 85344
    (928) 669-9211; www.crit-nsn.gov

Fort McDowell Yavapai Nation
    P.O. Box 17779, Fountain Hills, AZ 85269
    (480) 789-7111; www.ftmcdowell.org

Fort Mojave Tribe
    500 Merriman Avenue, Needles, CA 92363
    (760)-629-4591; mojaveindiantribe.com

Fort Yuma–Quechan Tribe
    P.O. Box 1899, Yuma, Az 85366
    (760) 572-0213

Gila River Indian Community
    P.O. Box 97, Sacaton, AZ 85247
    (520) 562-9840; www.gric.nsn.us

Havasupai Tribe
    P.O. Box 10, Supai, AZ 86435-0010
    (928) 448-2731; www.havasupaitribe.com

The Hopi Tribe
P.O. Box 123, Kykotsmovi, AZ 86039
(928) 734-2441; www.hopi.nsn.us

Hualapai Tribe
P.O. Box 179, Peach Springs, AZ 86434
(928) 769-2216; www.hualapai-nsn.gov

Kaibab Paiute Tribe
HC 65, Box 2, Fredonia, AZ 86022
(928) 643-7245; www.kaibabpaiute-nsn.gov

Navajo Nation
P.O. Drawer 9000, Window Rock, AZ 86515
(928) 871-6352; www.navajo.org

Pascua Yaqui Tribe
7474 S. Camino de Oeste, Tucson, AZ 85746
(520) 883-5000; www.pascuayaqui-nsn.gov

Salt River Pima-Maricopa Indian Community
10005 East Osborn Road, Scottsdale, AZ 85256
(480) 362-7400; www.srpmic-nsn.gov

San Carlos Apache Tribe
P.O. Box 0, San Carlos, AZ 85550
(928) 475-2361

San Juan Southern Paiute
Tuba City, AZ 86045
(928) 206-2648

Tohono O'odham Nation of Arizona
P.O. Box 837, Sells, AZ 85634
(520) 383-2028; www.tonation-nsn.gov

Tonto Apache Tribe
Tonto Apache Reservation #30, Payson, AZ 85541
(928) 474-5000; www.tontoapache.com

White Mountain Apache
P.O. Box 700, Whiteriver, AZ 85941
(928) 338-2500; www.wmat.nsn.us

Yavapai-Apache Nation
2400 W. Dubai, Camp Verde, AZ 86322
(928)-567-3649; www.yavapai-apache.org

Yavapai-Prescott Indian Tribe
530 E. Merritt St., Prescott, AZ 86301
(928) 445-8790

New Mexico

Jicarilla Apache Nation
P.O. Box 507, Dulce, NM 87528
(575) 759-3242; www.ausbcomp.com/redman/jicarilla.htm

Kewa Pueblo (formerly the Pueblo of Santo Domingo)
P.O. Box 99, Santo Domingo Pueblo, NM 87052
(505) 465-2214

Mescalero Apache Tribe
P.O. Box 227, Mescalero, NM 88340
(575) 464-4494

Navajo Nation
P.O. Box 7440, Window Rock, AZ 86515
(928) 871-6352; www.navajo.org

Ohkay Owingeh (formerly Pueblo of San Juan)
P.O. Box 1099, San Juan Pueblo, NM 87566
(505) 852-4400

Pueblo of Acoma
P.O. Box 309, Acoma, NM 87034
(505) 552-6604

Pueblo of Cochiti
P.O. Box 70, Cochiti, NM 87072
(505) 465-2244; www.pueblodecochiti.org

Pueblo of Isleta
P.O. Box 1270, Isleta, NM 87022
(505) 869-3111; www.isletapueblo.com

Pueblo of Jemez
P.O. Box 100, Jemez Pueblo, NM 87024
(575) 834-7359

Pueblo of Laguna
P.O. Box 194, Laguna, NM 87026
(505) 552-6654; www.lagunapueblo.org

Pueblo of Nambe
Route 1, Box 117-BB, Santa Fe, NM 87506
(505) 455-2038

Pueblo of Picuris
P.O. Box 127, Penasco, NM 87553
(575) 587-2519

Pueblo of Pojoaque
78 Cities of Gold Road, Santa Fe, NM 87506
(505) 455-3334

Pueblo of San Felipe
P.O. Box 4339, San Felipe Pueblo, NM 87001
(505) 867-3381

Pueblo of San Ildefonso
Route 5, Box 315-A, Santa Fe, NM 87506
(505) 455-2273

Pueblo of Sandia
481 Sandia Loop, Bernalillo, NM 87004
(505) 867-3317; www.sandiapueblo.nsn.us

Pueblo of Santa Ana
2 Dove Road, Santa Ana Pueblo, NM 87004
(505) 771-6701; www.santaana.org

Pueblo of Santa Clara
P.O. Box 580, Espanola, NM 87532
(505) 753-7330

Pueblo of Taos
P.O. Box 1846, Taos, NM 87571
(575) 758-9593; www.taospueblo.com

Pueblo of Tesuque
RR 42, Box 360-T, Santa Fe, NM 87506-2632
(505) 955-7732

Pueblo of Zia
135 Capitol Square Drive, Zia Pueblo, NM 87053-6013
(505) 867-3304

Pueblo of Zuni
1203B State Hwy 53; P.O. Box 339, Zuni, NM 87327
(505) 782-7000; (505) 782-7022; www.ashiwi.org

Oklahoma

Absentee-Shawnee Tribe of Indians of Oklahoma
2025 S. Gordon Cooper Drive, Shawnee, OK 74801
(405) 275-4030; www.astribe.com

Alabama-Quassarte Tribal Town
P.O. Box 187, Wetumka, OK 74883
(405) 452-3987; www.alabama-quassarte.org

Apache Tribe
P.O. Box 1220, Anadarko, OK 73005
(405) 247-9493

Caddo Nation of Oklahoma
P.O. Box 487, Binger, OK 73009
(405) 656-2344; www.caddonation-nsn.gov

Cherokee Nation
P.O. Box 948, Tahlequah, OK 74465
(918) 456-0671; www.cherokee.org

Cheyenne and Arapaho Tribes
P.O. Box 38, Concho, OK 73022
(405) 262-0345; www.c-a-tribes.org

Chickasaw Nation
P.O. Box 1548, Ada, OK 74821
(580) 436-2603; www.chickasaw.net

Choctaw Nation of Oklahoma
Drawer 1210, Durant, OK 74702
(580) 924-8280; www.choctawnation.com

Citizen Potawatomi Nation
1601 S. Gordon Cooper Drive, Shawnee, OK 74801
(405) 275-3121; www.potawatomi.org

Comanche Nation
P.O. Box 908, Lawton, OK 73502
(580) 492-3751; www.comanchenation.com

Delaware Nation
P.O. Box 825, Anadarko, OK 73005
(405) 247-2448; www.delawarenation.com

Delaware Tribe of Indians
170 N.E. Barbara, Bartlesville, OK 74006
(918) 336-5272; www.delawaretribe.org

Eastern Shawnee Tribe of Oklahoma
P.O. Box 350, Seneca, MO 64865
(918) 666-2435; www.estoo-nsn.gov

Euchee (Yuchi) Tribe of Indians
P.O. Box 10, Sapulpa, OK 74067
(918) 224-3065

Fort Sill Apache Tribe of Oklahoma
Rt. 2, Box 121, Apache, OK 73006
(580) 588-2298; www.fortsillapache.com

Iowa Tribe of Oklahoma
Rt. 1, Box 721, Perkins, OK 74059
(405) 547-2402; www.iowanation.org

Kaw Nation
Drawer 50, Kaw City, OK 74641
(580) 269-2552; www.kawnation.com

Kialegee Tribal Town
P.O. Box 332, Wetumka, OK 74883
(405) 452-3262; www.kialegeetribaltown.net

Kickapoo Tribe of Oklahoma
P.O. Box 70, McLoud, OK 74851
(405) 964-7065; www.kickapootribeofoklahoma.com

Kiowa Indian Tribe of Oklahoma
P.O. Box 369, Carnegie, OK 73015
West of Carnegie on Hwy 9
(580) 654-2300; www.kiowatribe.net

Miami Nation
P.O. Box 1326, Miami, OK 74355
(918) 542-1445; www.miamination.com

Modoc Tribe of Oklahoma
418 G Street S.E., Miami, OK 74354
(918) 542-1190; www.modoctribe.net

Muscogee (Creek) Nation
P.O. Box 580, Okmulgee, OK 74447
(918) 756-8700; www.muscogeecreeknation-nsn.gov

Osage Nation
813 Grandview, Pawhuska, OK 74056
(918) 287-5432; www.osagetribe.com.

Otoe-Missouria Tribe of Indians
8151 Highway 177, Red Rock, OK 74651-0348
(580) 723-4466; www.omtribe.org

Ottawa Tribe of Oklahoma
P.O. Box 110, Miami, OK 74355
(918) 540-1536; www.ottawatribe.org

Pawnee Nation
P.O. Box 470, Pawnee, OK 74058
(918) 726-3621; www.pawneenation.org

Peoria Tribe of Indians of Oklahoma
P.O. Box 1527, Miami, OK 74355
(918) 540-2535; www.peoriatribe.com

Ponca Nation
20 White Eagle Drive, Ponca City, OK 74601
(580) 762-8104; www.ponca.com

Quapaw Tribe
P.O. Box 765, Quapaw, OK 74363
(918) 542-1853; www.quapawtribe.com

Sac and Fox Nation
Rt. 2, Box 246, Stroud, OK 74079
(918) 968-3526; www.sacandfoxnation-nsn.gov

Seminole Nation
  P.O. Box 1498, Wewoka, OK 74884
  (405) 257-7200; www.seminolenation.com

Seneca-Cayuga Tribe
  P.O. Box 1283, Miami, OK 74355
  (918) 542-6609; www.sctribe.com

Shawnee Tribe
  P.O. Box 189, Miami, OK 74355
  (918) 542-2441; www.shawnee-tribe.com

Thlopthlocco Tribal Town
  P.O. Box 188, Okemah, OK 74859-0188
  (918) 560-6198

Tonkawa Tribe
  1 Rush Buffalo Road, Tonkawa, OK 74653
  (580) 628-2561; www.tonkawatribe.com

United Keetoowah Band of Cherokees
  P.O. Box 746, Tahlequah, OK 74465
  (918) 431-1818; www.keetoowahcherokee.org

Wichita and Affiliated Tribes (Wichita, Keechi, Waco, and Tawakonie)
  P.O. Box 729, Anadarko, OK 73005
  (405) 247-2425; www.wichitatribe.com

Wyandotte Nation
  64700 East Highway 60, Wyandotte, OK 74370
  (918) 678-2297; www.wyandotte-nation.org

Texas

Alabama-Coushatta Tribes of Texas
  571 State Park Road, Livingston, TX 77351
  (936) 563-1100; www.alabama-coushatta.com

Kickapoo Traditional Tribe of Texas
  HC 1, Box 9700, Eagle Pass, TX 78852
  (830) 773-2105; www.ktttribe.org

Lipan Apache Tribe of Texas
  P.O Box 8888, Corpus Christi, TX 78468
  (361) 215-5121; www.lipanapache.org

Ysleta del Sur Pueblo of Texas (Tigua)
  P.O. Box 17579, Ysleta Station, El Paso, TX 79917
  (915) 859-8053; www.ysletadelsurpueblo.org

SELECTED REFERENCES

Alvarez, Elizabeth Cruce, ed. *Texas Almanac 2004–2005.* Dallas: Dallas Morning News, 2004.

Barnes, Will C. *Arizona Place Names.* Tucson: University of Arizona Press, 1935.

Beck, Warren A., and Ynez D. Haase. *Historical Atlas of New Mexico.* Norman: University of Oklahoma Press, 1969.

Brian, Nancy. *River to Rim: A Guide to Place Names along the Colorado River in Grand Canyon from Lake Powell to Lake Mead.* Flagstaff, Ariz.: Earthquest, 1992.

Bright, William. *Native American Placenames of the United States.* Norman: University of Oklahoma Press, 2004.

Debo, Angie. *And Still the Waters Run: The Betrayal of the Five Civilized Tribes.* Princeton, N.J.: Princeton University Press, 1940. Reprint, Norman: University of Oklahoma Press, 1984.

Gibson, Arrell M. *Oklahoma: A History of Five Centuries.* 2nd edition. Norman: University of Oklahoma Press, 1981.

Goodman, James. *The Navajo Atlas: Environment, Resources, Peoples, and History of the Diné Bikeeyah.* Norman: University of Oklahoma Press, 1987.

Granger, Byrd Howell. *Arizona's Names.* Tucson: Treasure Chest, 1983.

Holland, C. Joe. *A Pronunciation Guide to Oklahoma Place Names.* Norman: University of Oklahoma School of Journalism, 1950.

Julyan, Robert. *The Place Names of New Mexico.* Albuquerque: University of New Mexico Press, 1998.

Linford, Laurance D. *Navajo Places: History, Legend, Landscape.* Salt Lake City: University of Utah Press, 2000.

Martin, Jack B., and Margaret McKane Mauldin. *A Dictionary of Creek/Muskogee, with Notes on the Florida and Oklahoma Seminole Dialects of Creek.* Lincoln: University of Nebraska Press, 2000.

Mithun, Marianne. *The Languages of Native North America.* New York: Cambridge University Press, 1999.

Morris, John W., and Charles Goins. *Historical Atlas of Oklahoma.* Norman: University of Oklahoma Press, 1986.

Read, William Alexander. *Florida Place-names of Indian Origin and Seminole Personal Names*. Baton Rouge: Louisiana State University Press, 1934.

Sheridan, Thomas E. *Arizona: A History*. Tucson: University of Arizona Press, 1995.

Shirk, George H. *Oklahoma Placenames*. Norman: University of Oklahoma Press, 1974.

Stephens, A. Ray, and William M. Holmes. *Historical Atlas of Texas*. Norman: University of Oklahoma Press, 1989.

Tarpley, Fred. *Texas Place Names*. Austin: University of Texas Press, 1980.

Texas State Historical Society. *Handbook of Texas Online*. www.tshaonline.org.

U.S. Board of Geographic Names. www.geonames.usgs.gov/domestic.

Walker, Henry P., and Don Bufkin. *Historical Atlas of Arizona*. 2nd edition. Norman: University of Oklahoma Press, 1986.

Wilson, Alan. *Navajo Place Names*. Guilford, Conn.: Jeffrey Norton, 1995.

Wright, Muriel. *A Guide to the Indian Tribes of Oklahoma*. Norman: University of Oklahoma Press, 1986.

———. "Some Geographic Names of French Origin in Oklahoma." *Chronicles of Oklahoma*, vol. 7, no. 2 (June, 1929), pp. 188-93.

Young, Robert W., and William Morgan. *The Navajo Language: A Grammar and Colloquial Dictionary*. 2nd edition. Albuquerque: University of New Mexico Press, 1987.

ACKNOWLEDGMENTS

Some of the following persons assisted William Bright with information for his 2004 book, *Native American Placenames of the United States*, and others he intended to thank for help with this volume. Each is either a recognized specialist in a particular language or language family or a fellow placenames scholar whose work or advice he drew on. They are James Armagost, Mount Vernon, Washington (Comanche); Irvine Davis, Albuquerque, New Mexico; T. Wayne Furr, Oklahoma Board on Geographic Names, Norman (Oklahoma placenames); Philip Greenfield, San Diego State University, California (Apachean languages); Jane H. Hill, University of Arizona, Tucson (Tohono O'odham); Kenneth C. Hill, Tucson, Arizona (Hopi, O'odham); Leanne Hinton, University of California, Berkeley (Havasupai); Robert Julyan, New Mexico Board on Geographic Names, Albuquerque (New Mexico placenames); Jack Martin, College of William and Mary, Williamsburg, Virginia (Muscogee); John McLaughlin, Utah State University, Logan (Comanche); Pamela Munro, University of California, Los Angeles (Mojave, Choctaw, Chickasaw); Tim Norton, Arizona State Placename Board, Phoenix (Arizona placenames); Robert Rankin, University of Kansas, Lawrence (Siouan languages); Willem de Reuse, University of North Texas, Denton (Apachean languages); David Rood, University of Colorado, Boulder (Caddoan languages); Emory Sekaquaptewa, Hopi Tribe, Arizona (Hopi); Laurel Watkins (Kiowa); Adrian Wilson, Gallup, New Mexico (Navajo); Yukihiro Yumitani, Sanyo Gakuen University, Japan; and Ofelia Zepeda, University of Arizona, Tucson (Tohono O'odham).

The editors, who drew on our own knowledge to fill in some of Bright's incomplete entries (O'Neill on Athabaskan languages, Anderton on Caddo, Comanche, Euchee, and Ponca), would also like to thank the following people, who helped in their particular areas of expertise: Geneva Navarro, Albuquerque, New Mexico (Comanche); Gus Palmer, University of Oklahoma (Kiowa); and Jim Rementer, Lenape Language Project, Delaware Tribe of Indians, Bartlesville, Oklahoma (Delaware/Lenape).

www.ingramcontent.com/pod-product-compliance
Lightning Source LLC
Chambersburg PA
CBHW021331090426
42742CB00008B/564